Truth in Science, the Humanities and Religion

BALZAN SYMPOSIUM 2008

TRUTH IN SCIENCE, THE HUMANITIES AND RELIGION

organized by the
International Balzan Foundation
at the
University of Lugano (Università della Svizzera italiana)
16–17 May 2008

PROCEEDINGS

edited by

M.E.H. NICOLETTE MOUT
AND WERNER STAUFFACHER

Springer

Editors
M.E.H. Nicolette Mout
Piazzetta U. Giordano, 4
20122 Milano
Italy
balzan@balzan.it

Werner Stauffacher
Piazzetta U. Giordano, 4
20122 Milano
Italy
balzan@balzan.it

ISBN 978-94-017-8132-9 ISBN 978-1-4020-9896-3 (eBook)
DOI 10.1007/978-1-4020-9896-3
Springer Dordrecht Heidelberg London New York

Printed on acid-free paper

Springer is part of Springer Science+Business Media (www.springer.com)

Foreword

Since 1961, the International Balzan Foundation awards prizes for outstanding achievements through its General Prize Committee. A prize for Humanity, Peace and Brotherhood among peoples is awarded, at an interval of not less than three years to personalities and institutions for and in support of their contribution to humanitarian work, brotherhood and peace in the world. In addition, two prizes each are awarded every year for scholarly excellence in the Humanities – i.e. the arts, humanities and social sciences – and for fundamental research in the Sciences – i.e. physics, astronomy, mathematics, natural sciences and medicine together with epidemiology. The four fields to be awarded – two in the Humanities, and two in the Sciences – change every year. No more than four prizes may be awarded in any one year. The General Prize Committee takes care to focus on those disciplines which are not covered by other well-known international prizes.

More recently, the Foundation has begun to organize symposia about themes of overriding significance and importance, both for the scientific and scholarly community and for the general public, with the intention to encourage and intensify interdisciplinary discussions and debate among scientists and scholars from vastly differing horizons, but also inviting personalities from the "outside-world" of politics and economy. Thus, in 1988, following the surge of ethical awareness and resulting constraints in science, biology and medicine, an international seminar on "Challenges of Bioethics" was held in Venice. In 2002, a symposium followed at the Royal Society in London. Under the heading of "Meeting the Challenges of the Future", it was devoted to the dialogue between "the two cultures", i.e. between the Humanities on one hand, and the Sciences on the other.

For 2008, the theme chosen by the General Prize Committee was "Truth". Truth in the humanities and the sciences as well as in religion and theology; truth, not in contrast to fraud but as the ultimate goal of scientific endeavour of any kind, the truth that is said to be "the horizon of scientists and philosophers". A small preparatory committee of members of the General Prize Committee was asked to draw up the program. The Balzan Foundation invited 21 distinguished scholars and scientists – several of them Balzan Prize Winners – as speakers and panel members as well as a selected audience of about 200 academics and scholars from mostly European Universities, academic Institutions and learned Societies – among which a number of students from Italian and Swiss universities selected by the Academies of the two countries – to the Italian-Swiss University of Lugano for the two-day-Symposium.

The question of the meaning of "truth" is central to many areas of contemporary debate, whether between those subscribing to a post-Enlightenment view of the world and those who seek fundamental truth in religious texts, or between those maintaining that there are absolute truths and those believing facts to be social constructs. For some, the ultimate truth is revealed through religious faith and textual authority. Can this view be reconciled with an evidence-based, materialist, post-Enlightenment perspective of the truth as embraced by the natural sciences? If religion holds the key to the truth, which religion and which truths?

During the five thematic sessions of the symposium, all attended by the same audience and by all the speakers and panel members, these and many other questions, but in particular the one about the meaning of truth, were examined and debated. The whole range of perspectives represented on the panels and in the audience came to the fore. After the keynote lecture by Professor Simon Blackburn, the five sessions covered the following disciplines: philosophy, mathematics, physics, cosmology, the biological sciences including biodiversity and sustainability, history, the social sciences, theology and religion.

This volume contains the speakers' lectures, the corresponding reactions of the invited panel members as well as the panel and general discussions of the two-day-symposium. Thus the interested reader can familiarize himself with views ranging from philosophy to religion, from mathematics, physics and cosmology to history and the social and natural sciences. Undoubtedly, he will discover conflicts and discrepancies but also lines of convergence. In doing so, he will find that the adjectives which – perhaps unfortunately – are most frequently associated with the word "truth" are "provisional" or "relative". In Session VI, finally, the reader will find the concluding impressions of two eminent participants who were instrumental in the conception of the symposium, Professor Simon Blackburn, author of the book *Truth*, and Lord Krebs, member of the preparatory committee, who suggested the title of the symposium after having read the book. Rather than attempting to summarize the two days of conferences and debate, they put their fingers on critical moments of the discussions where discrepancies, possible convergences, but also possible avenues for future research and reflection seemed to come to light. Unless a reader is specifically interested in one particular viewpoint and immediately chooses to read the corresponding chapter, one might suggest to him to begin his reading with Blackburn's and Krebs' succinct and enlightening retrospective analysis of the symposium.

M.E.H. Nicolette Mout and Werner Stauffacher

Contents

Opening Session

16 May 2008 – 11:00 a.m.

Addresses of Welcome:
Giorgio Giudici, Mayor of Lugano
Piero Martinoli, President of the University of Lugano
Bruno Bottai, Chairman of the Board of the International Balzan "Prize" Foundation

Introduction and Chair:
Sergio Romano, Chairman of the General Prize Committee of the International Balzan "Prize" Foundation

Keynote Lecture:
Truth and Ourselves: the Elusive Last Word

Simon Blackburn

Sergio Romano:

Welcome to our meeting, in a city which is very Italian and very Swiss at the same time. In fact, it has the extraordinary characteristic of being what one would call an ambiguity or, as others would call it, an extraordinary synthesis. This city is our host, and in a certain sense it is a symbol of the Balzan Foundation, which is an Italian-Swiss foundation.

I don't think we could hold our conference in a better place, especially because we are in this city's university. Only three decades ago it did not exist, and now it has assumed the role of being one of the most interesting among the Italian language universities. Because of its young age its fields of activity are maybe still somewhat limited, but they attain a very high level.

Giorgio Giudici:

I am particularly proud of welcoming you, on behalf of the city of Lugano, to this important meeting promoted by the International Balzan Foundation. The Foundation has requested the presence of famous scholars in order to reflect on such an important theme as truth, a theme underlying all our actions, all our research, as well as

Balzan (ed.), *Truth in Science, the Humanities and Religion: Balzan Symposium 2008*,
DOI 10.1007/978-1-4020-9896-3_1, © Springer Science+Business Media B.V. 2010

our decisions in the field of politics and of lawgiving. When we ask ourselves whether what we are doing is right or not, we also address the question whether a particular truth is valid in a moral context. In this sense, we know that there is no absolute and universally shared truth. Indeed, what is true in the universe of science may not be true, or may not be recognized as such, in the universe of morality. There are evident examples supporting this in the field of science. This is why I am inviting people to follow reason. As Schopenhauer said, those who appreciate truth should also listen to good reasons even when they come from the mouth of their opponents. Nor should we ever forget in our disputes what Voltaire said, namely that peace is worth more than truth. The fact that many participants of very high scholarly standing have come to Lugano to discuss these themes, and the fact that these themes will be discussed in an interdisciplinary manner point to the best way to understand the theme itself. This is why I am particularly grateful to the Balzan Foundation for choosing our city and this university to promote reflection on a value of such high importance.

The International Balzan Foundation and this conference bring new lustre to this recently built university. It was built at this very site and inaugurated in October 1996 in accordance with the idea of bringing academia to the very heart of the city. This has helped to make up for the deficiency that had relegated Lugano, one of the important cities in Switzerland, to a marginal role for a long time in recent history. Lugano now intends to play a more important role in the society of knowledge, and it aspires to be – and the conditions for this are right – to become a crossroads of communication and ideas. Indeed, because of the foundation of this university, together with the new important programmes on which we are working in order to complete and broaden the spectrum of this new cultural centre, it makes sense to envisage Lugano as a centre and a reference point for cultural and economic relations and exchanges between the North and the South of Europe. This important conference enhances our hope that we can achieve this aim.

Piero Martinoli:

I am happy to welcome you on behalf of the Italian-Swiss University.

I think I reflect the opinion of all my colleagues in our faculties of Architecture, Sciences, Communication, Information technologies and Economics in saying that I am proud that our university is hosting the Balzan Foundation for this two-day conference about 'Truth'. Ever since I became familiar with the topics on the agenda, I realized that this meeting would be extremely interesting and that it is fully in line with both the missions of our university and of the Balzan Foundation. Thirty research institutes, twenty two-year Masters' programmes, and many hundreds of university students – these are just some of the figures reflecting how young and dynamic our academic universe is. It brings together very different people from varying national and social backgrounds and reconciles topics as far apart as the study of financial systems and that of mass media. Here, our students find a very fertile ground promoting intellectual exchange and confrontation in dialectics. While promoting a multidisciplinary approach, we were able to identify new values stemming from the recent history of our university and to incorporate them in its guidelines for the near future.

During the press conference introducing this symposium, Professor Veca has made some very interesting remarks about multidisciplinarity as a founding trait of the Balzan Foundation. He called it the Foundation's DNA, and also its test field for more open approaches to academic culture. These approaches are no longer perceived as ensuing from narrow specializations within the context of a well-fixed discipline, but rather as the result of a plurality of perceptions and understanding within very different areas of knowledge.

In order to be creative and productive, this idea of academic culture requires from university students and university professors a very open mindset, goodwill, and also some degree of humbleness. They will have to explore what's going on in other countries and to look into the way different areas of knowledge interact. All the while they are supported by the belief that new knowledge and new discoveries can only thrive if a multidisciplinary approach is taken. This multidisciplinary approach, which we share with the Balzan Foundation, is an extremely important trait in our academic culture, and I do believe that by adopting it wholeheartedly we have really set out on the right path. We want to promote learning and research, starting from a very broad approach to knowledge and culture. There are still hidden truths in our time, and to discover them, one needs to embark on a new approach. In the future we will need to move more and more away from the reflected images of Plato's cave. We will need to look into the sunlight and to start considering, with the help of a new vision, a new universe, still unexplored: a multidisciplinary universe where medicine and ethics, informatics, information technology and biology are all intertwined.

This approach probably initiates a new chapter in the history of knowledge, and I'm very happy that I can share this meeting, which is a true example of it, with you – philosophers, intellectuals, scientists, free spirits, distinguished experts from different backgrounds. When pursuing the truth it is really necessary to focus on quality, on dialogue, on plurality, and as I have mentioned before, you also need to be humble. I hope this conference will bring some light.

Bruno Bottai:

The Balzan Foundation is very happy to see this conference begin and to have this happen here, in the Ticino. I have been its president for a number of years, and during that time I have tried to create links with the Ticino. It is just a few kilometres away from Milan, and, by being part of both *italianità* and polyglot Switzerland, it gives Italian culture the privilege of being a constitutive part of the polyglot Helvetic Confederation to which Italy is linked, not only through language and culture, but also through a chain of historical events common to both countries.

I am happy to thank the Mayor of Lugano and the Rector of the University. In addition, I should also like to thank Mrs. Suzanne Werder, Secretary General of the Balzan "Prize" Foundation, and the Consul General of Italy, whose presence here today is a great honour for our Foundation. I would also like to thank Sergio Romano, the ever-active Chairman of the Foundation's General Prize Committee who contributes such a great deal to the success our foundation in Milan, and who, as chairman of the Committee, was in charge of the preparation of this event.

It is a great pleasure for the foundation to see so many students, so many illustrious academics from so varying fields of science, the humanities and theology meeting here in Lugano. I consider it extremely important that there should be dialogue, not just discussion, between different cultures and identities. This is also the reason why the General Prize Committee, which is made up of academics coming from all the important disciplines and originating from eleven different European countries, try to promote an interdisciplinary dialogue. The Balzan "Prize" Foundation confers awards annually. These are significant prizes, both in terms of their international renown and their financial importance. They are destined to stimulate the international scientific and scholarly culture. Normally, two prizes are awarded annually for the humanities and two for the natural sciences.

In conclusion, I should like to draw your attention to the fact that this conference is attended, on one hand, by a large number of young academics from the University of Lugano, and on the other, by students from the universities of Basel, Geneva, St. Gallen, Zurich, Como, Milan, Pavia, Pisa and Rome. They have been selected and invited by the Italian National Academy of Lincei, and by the association of the Swiss Academies of Arts and Sciences. This emphasizes both the academic inspiration and the Italian-Swiss binational character of the Balzan Foundation.

Sergio Romano:

The Balzan Foundation has a quite recent tradition of convening symposia. It awards four prizes for academic achievements together with a prize for peace, solidarity and brotherhood, and every four or five years it organises meetings on a specific relevant topic. Our task is not to run after red-hot topics, things you can read about in the mass media, but nevertheless we are keenly interested in what really matters nowadays. Today, therefore, we will be discussing a topic which is extremely important to all of us, in both the academic and the religious world. In recent times, we have been confronted with new problems and these are challenging both scientists and philosophers. This is why we thought that organising a conference on truth could be extremely interesting.

Let me now introduce the first speaker, Simon Blackburn, philosopher from the University of Cambridge. His introductory lecture will deal with the topics we are going to focus on. Just visit his website, and there you will see how a philosopher is able to introduce himself and his ideas in a very clear and attractive way, also for laymen, without losing the depth of his thought.

Simon Blackburn:

It's a great honour to be introducing this symposium. I apologize that I have to speak in English – if I tried to speak in Italian, it would be a much less rewarding experience for all of you.

Keynote Lecture

Truth and Ourselves: the Elusive Last Word

Simon Blackburn:

We academics like to regard ourselves as the guardians. We are the profession especially charged with sifting fact from fiction, with testing opinion and conducting investigation. Opinions are born into a Darwinian jungle, and it is our special task to ensure that only the fittest survive. However, unlike biological evolution, the process, we believe, has an upward direction. Later opinions will tend to be free of infirmities that may have infected earlier ones. Unlike biological evolution, the process may also have an end, or so some people think. The end of the process will be the truth, and even now we are much nearer to it than we were at the beginning. At times, as the process unfolds, there may be checks and stoppages, even local minima when opinion gets into a trough from which it cannot easily recover. But these, we hope and believe, are at worst temporary setbacks. Overall, the process has a direction, and we are proud to help to push it on its journey.

This is what we like to think. But of course there have always been sceptics. Sometimes scepticism is local. We may wonder whether there is such a thing as truth in the domain of ethics, for example, or in the domain of taste, in activities of interpretation, or in history. Such local scepticism is often contrastive. It is because ethics, or judgements of taste, or history, are found to lack some special mark of more respectable inquiry that they become objects of suspicion. Scientists, I suspect, are apt to believe that their own disciplines carry the special mark, whereas the humanities, or softer sciences, do not. But if we are comfortable on some rung of this ladder, we might do well to reflect whether the hierarchy itself has a certificate of truth, or whether it may reflect the ideology of our particular time and historical moment. After all, historically there is no category of human opinion that has escaped the cold winds of scepticism. The evidence of the senses, the building blocks of scientific theory, mathematics and logic themselves, not to mention our favourite stories about ourselves, our histories, our ethics and our gods, have all at various times found themselves embarrassed by the cold light of reason.

Scepticism, as this reminds us, has many forms, and a long philosophical history. But so do answers to scepticism. One famous answer was given by a predecessor of mine in my Chair in Cambridge, G. E. Moore. Moore's strategy was to take a good, common-sense proposition from the local area, and ask us to compare our commitment to it with our confidence in whichever philosophy of the matter we were using to cast doubt upon its fitness for truth. Suppose, for example, ethics has come under sceptical suspicion. Nevertheless, are we not certain that, for instance, it is a bad thing that human beings find it easy to be cruel to each other, or simply indifferent to the sufferings of others? Are we not *more* certain of that than we are of whatever subtle philosophical reflection led us to worry about its fitness for truth? If historical truth, with its mutable categories and plural perspectives comes under attack, well, are we not certain that, as Carlyle put it, "Hunger and nakedness and righteous oppression lying heavy on twenty-five million hearts" had something to contribute to the French revolution? Might we not even attach more confidence to Carlyle's generalization that this will also be a necessary condition of all future revolutions, than we do to whatever metaphysics or epistemology led us into the corrosive paths of scepticism? And so on for whichever domain we find ourselves in doubt about.

Moore's strategy – which has a pedigree as old as Aristotle – can be effective, and many philosophers have commended it. But it may leave us intellectually dissatisfied, depending on the grip that sceptical arguments held on us in the first place. If we are unable to shake off their grip, then we may feel not so much reassured by Moore, as trapped in some kind of inescapable *aporia*. We will think of ourselves as victims, forced by our human nature to distribute confidence here and there, but unable by reason to certify or authenticate such confidence as we have. For example, surveying the variety of theological or moral opinions that have commended themselves at different places and times, we may gravitate towards the sceptical view that no one of them is better than any other, and hence that none deserves any confidence whatsoever. And then if Moore teases out the confession that nevertheless we cannot shake off some theological or moral conviction that we happen to have acquired, we may feel intellectually guilty or ashamed. Reason, as David Hume put it, has its privileges as well as natural belief. If we cannot see our propensities to form some beliefs as themselves selecting for the truth, then the result must be a 'melancholy' confusion, a fragmented self that finds itself saying things that it also says do not deserve the confidence temporary suspension of confidence or inability to find our voice.

Scepticism in modern Europe is often traced back to Montaigne and the rediscovery of the classical tradition, particularly in Sextus Empiricus. It rumbled on in Descartes and Hume. But the most potent contemporary image derives from Kant. Kant gave Europe the idea that the cognitive system imposes its own order on the manifold: we never come across raw, uncontaminated data. The most basic experience will have been processed, synthesized, categorized, or, as William James summed it up, *cooked* for us to digest it. Kant thought that the process was universal, and that, provided we accepted a degree of modesty, he could provide an a priori pedigree for it, warding off the sceptical threat. But it was not long before Europe listened to Hegel, and the cooking process was thought of as contextualized, historicized, made contingent upon the moment we happen to occupy, but very likely invisible to agents at the time. Nor was this pessimism confined to philosophy, or unaffected by wider cultural currents. The collapse of any defensible faith in a providential harmony between mind and world gained momentum in the mid nineteenth century with the collapsing historicity of the Bible. Darwin accelerated the process by introducing the pragmatic note, suggesting that what matters in the life of an animal, including the human animal, was not whether it represented the world as it is, but simply whether it works, whether it functions. The first philosopher to take the measure of this was Nietzsche:

> Here and there we [...] laugh at the way in which precisely science at its best seeks most to keep us in this *simplified*, thoroughly artificial, suitably constructed and suitably falsified world – at the way in which, willy-nilly, it loves error, because, being alive, it loves life.[1]

And hence:

> Whatever philosophical standpoint one may adopt today, from every point of view the *erroneousness* of the world in which we think we live is the surest and firmest fact that we can lay eyes on.[2]

In the contemporary world, we do not hear so much about scepticism as relativism, and sometimes social constructivism, although I shall talk mainly in terms of the former. The difference between relativism and scepticism is that while for the sceptic there is truth – good, objective, truth – it is difficult, and perhaps impossible, for us to obtain it. Relativism by contrast does without the ideal, and instead finds too many truths: your truth, his truth, my truth. Relativism is, as it were, scepticism for democrats. For classical scepticism, the result was to give up conviction and certainty; for their relativist descendants, you can have as much certainty as you like, for all opinions stand alike as the mere reactions of some persons, calibrated some ways, at some times.

Relativism is bound to be the especial, viscerally hated enemy of those claiming power and authority, in science or any other part of life. As we may remember, on the eve of his election to the Throne of St Peter, Pope Benedict XVI preached not against poverty or war or greed, but against relativism. The problem is that the relativist hears such preaching in his own way, and the conservative is powerless to prevent that. In my book *Being Good* I tell a (true) story about this, and at the risk of boring anyone who has read that book, I would like to tell it again here. It concerns a friend of mine, who had a Fellowship at a high-powered ethics institute, which had put on a forum in which representatives of the great religions held a panel. First the Buddhist talked of the ways to calm, the mastery of desire, the eightfold path of enlightenment, and the four noble truths, and the panellists were all very appreciative, and said "Wow, terrific, if that works for you that's great". Then the Hindu talked of the cycles of suffering and birth and rebirth, the teachings of Krishna and the way to release, and again they all said "Wow, terrific, if that works for you that's great". And so on, until the Catholic priest talked of the Fall of man, the message of Jesus Christ, the promise of salvation and the way to life eternal, and they all said "Wow, terrific, if that works for you that's great". And he thumped the table and shouted: "No! It's not a question of if it works for me! It's the true word of the living God, and if you don't believe it you're all damned to Hell!"

And they all said: "Wow, terrific, if that works for you that's great".

The reason I like this story is that it illustrates a certain kind of impasse philosophers meet when they try to defend what I called the conservative view that there are real standards, and that even in ethics and aesthetics, history or sociology, we can aspire

[1]Nietzsche, *Beyond Good and Evil* II, §24.
[2]Nietzsche, *Beyond Good and Evil* II, § 34.

to opinions that are objective and reasonable, and even *true*. The opponent to fear is not so much someone who straightforwardly disagrees with some judgement of ours in these areas. We are on all fours with *him*. We point out the features of whatever it is we are talking about. We think that these features ought to sway him, and he points out the ones that, he thinks, ought to sway us, and this goes on until one of us is swayed, or we agree to differ, or we go to war. The opponent that irritates us is rather someone who conceives of himself as above the fray altogether. He has achieved a kind of God's eye view of the debates that embroil lesser mortals. He sees the operations of the human kind for what they are, the historical evolutions of animals trying to get by, the various readings of differently calibrated instruments, the clashes of wills. He sees ploys and manipulations and exercises of power. He sees persuasion and spin. But if any of the participants start talking of reason and truth, he sees only the same again, with added decorations.

This cynical or disengaged character is not new, of course. He is the principal opponent of Socrates in many of the *Dialogues*, and it took all Plato's powers of reasoning to suppress these Gorgiases and Protagorases and Callicleses. I nearly said, all Plato's powers of persuasion, for one of the more obvious targets for deconstruction in the *Dialogues* is the fact that mostly Plato really only manages to fight spin with spin. To unseat rhetoric and substitute philosophy, he offers us not philosophy, but the rhetoric of the Cave and the Line, and the flattering self-image of the philosopher ascending to contact with the true norms: the Forms, or guarantors of objectivity and truth. But this is a vision, and a cloudy one at that, rather than an argument.

However, Plato did invent another argument, the so-called *peritrope*. The idea is that sceptics and relativists are in some kind of self-contradiction. The idea is first to get them to assert something. So, for example, the relativist might say: "Truth has been unmasked as the rhetoric of those making a move in a power-game. So, all opinions are equally good or equally true". The Platonist now seizes on the consequence: "You say that all moral opinions are equally good. But then *my* opinion is that some are better than others. So that is true for me; and after all your opinion is only at best true for you, for by your own principles you cannot advance it as true absolutely."

This argument has been echoed by a great many eminent, and otherwise divergent, modern philosophers, including Bernard Williams, Hilary Putnam, and Thomas Nagel. A pointed application of it is the familiar problem with Marxism. Marx thinks that all values are ideology: that is, they are the result of particular economic forces at particular times, and have no further trans-historical authority. But he also *holds* various values, for instance that the world in which communism reigns will be intrinsically fairer and more just than the world under capitalism. There is a strain here. If this opinion is itself just a piece of ideology, as it appears it has to be, then how should Marx himself take it seriously? If he is aware that this is just the kind of steam that you would expect from some classes of the intelligentsia at the particular socio-economic juncture represented by mid-nineteenth-century capitalism, must this not dilute and eventually destroy his enthusiasm for his cause? The problem is that the relativist or postmodernist or sceptic has to stand somewhere, while all the time seeming to deny that there is anywhere to stand.

It looks as though this trouble will arise whenever the doctrine will fall within its own scope. Is the relativist then in conflict with himself, needing both to advance the fateful claim – all opinions are equally good – as really, absolutely, true, and yet to accept at the same time that it is at best modestly true, true only for him, now?

I like the reply given by William James at this point. James calls the people agreeing with his own pragmatist downgrading of truth not relativists, but humanists, and writes:

> But can there be self-stultification in urging any account whatever of truth? Can the definition ever contradict the deed? "Truth is what I feel like saying" – suppose that to be the definition. "Well, I feel like saying that and I want you to feel like saying it, and shall continue to say it until I get you to agree". Where is there any contradiction? Whatever truth may be said to be, that is the kind of truth which the saying can be held to carry. The *temper* which a saying may comport is an extra-logical matter. It may indeed be hotter in some individual absolutist than in a humanist, but it need not be so in another. And the humanist, for his part, is perfectly consistent in compassing sea and land to make one proselyte, if his nature be enthusiastic enough.[3]

In James's view a Protagoras who buttonholes you and insists and insists that truth is relative is in no kind of self-contradiction. He wants to say this, and he wants you to say it as well, and he is enthusiastic enough to find the enterprise of converting you worthwhile. He is an assertive relativist.

James thinks the relativist can well take a stand, because taking a stand is not in itself an activity he opposes. All he opposes is what he sees as a particular high-flown and delusory conception of truth in which some people clothe the stands they take. Such people think you cannot have the activity of taking a stand without the clothing, which is why they find the relativist suspicious. The absolutist thinks that the dignities of truth and reason are necessary clothing, robes of State without which our activities could not take place, for they need these dignities. We must see ourselves as lovers of the truth, and servants of the good, so we must work to make the robes of truth, absolute truth, fit us properly. But where others see robes of state, the relativist sees only masks and disguises. But he happily mingles with the carnival, wearing the mask and disguise he has found on other people.

As usual in philosophy, when the sides line up so quickly, we might want to worry whether there is a shared mistake each is making. Here the obvious common factor is the view that truth and its handmaidens are indeed *clothing*, whether masks or robes. But suppose they are not? Less metaphorically, suppose there is nothing in the notion of truth, or even absolute truth, to be contested. Suppose truth is, as it were, too *small* to sustain the battle?

Many contemporary philosophers of logic and truth think this way, following Frege, Wittgenstein, Ramsey and Quine. They go by the name of deflationists or minimalists. The details vary, but the common starting point is a crucial observation, which is that if we consider any proposition p there is no difference between asserting, believing, denying, considering or pondering whether it is true that p and asserting, believing, denying, or pondering p itself. Truth is transparent, or invisible. Deflationists think that this is the key to understanding the whole concept.

[3]James, *The Essential Writings*, 270.

Why does this make a difference to the debate? The idea is that the battle between the conservative and the relativist is sustained by a conviction that there are two issues, when in fact there is only one. For this battle to be joined, there must first be some ordinary issue. Let us take the ethical issue of whether capital punishment should be allowed. But second there is a further, philosophical or reflective (second-order) issue: whether there is a truth of the matter, that capital punishment should be allowed. The conservative says yes, at the cost of buying into a mysterious world of normative reality; the relativist will not pay this cost, and so says no. The deflationist however denies that there is any second-order issue. There is only the issue of whether capital punishment should be allowed, full stop. If we hammer this out, and decide that it should be, then we do not increase the theoretical temperature by adding "what's more, that's true". This just repeats the conclusion that capital punishment should (or should not) be allowed, a conclusion to the moral issue with which we started.

Each side in the original debate got something right, and something wrong. The absolutist or conservative was right that it takes some work to decide whether capital punishment should be allowed. It requires thinking hard about such matters as revenge, deterrence, the powers of the State, and these are confusing and contested concepts. It is not a judgement that ought to come tripping off the tongue. The relativist was right that anyone's verdict on such a matter will be a function of a good many factors in their background, including for example recent history, and other determinants of admiration, disgust, shame and pride. But neither of them was right that second-order reflection, focused on worries about truth, either precedes, or helps or hinders decision about whether capital punishment should be allowed. To come to an opinion about that, your gaze must be firmly fixed on capital punishment, and the confused factors suggesting either that it is, or is not, a process to advocate. And once you have thought about such matter, you may be inclined to take a stand. You might then say "capital punishment should not be allowed", or, for it is the same thing: it is true that capital punishment should not be allowed. Or: it is really true, a fact, that capital punishment should not be allowed. Or: capital punishment should not be allowed; you had better believe it. Or: it corresponds to the eternal normative order that capital punishment should not be allowed. For none of these say any more than what you started with. We might have thought there was an ascending ladder here. But the ladder – which I like to call Ramsey's ladder, in honour of one of the first great deflationists, Frank Ramsey of Cambridge – is horizontal. It takes you nowhere.

But isn't that what the relativist said all along? And how has deflationism helped the conservative half of us? The conservative is going to feel short-changed. He thundered out the need for standards, real standards, and the philosopher has offered him only some very pallid version of what he wanted. He wanted truth, and objectivity, or the ability to see truth wherever it stands. The philosopher says he has given him it, but it wasn't much to give. Wherever you have an opinion, says the deflationist, you can express it, or you can express it with "it is true that…" prefaced to it. It makes no odds.

Conservatives who want more have tried to unseat deflationism. One argument is that we need to retain the idea of truth as a *norm* or ideal of inquiry. But the deflationist will reply that he can perfectly well retain whatever is valuable in that idea. It is a way, he says of generalizing over a mass of claims of this form: you ought to believe that there is a cat in the garden if and only if there is a cat in the garden. You ought to believe that every even number is the sum of two primes if and only if every even number is the sum of two primes. Such claims can be stated without any use of a notion of truth, but we sum up the whole mass of them by saying that you ought to believe what is true. Truth is nothing but a device of generalization. Similarly, the conservative may use what Hilary Putnam and Richard Boyd advanced as a "no miracles" argument for truth. If science did not have some substantial success to its credit, some title to be representing the world as it is, it would be a miracle that it is so successful. To which the deflationist again replies that if you give him a scientific claim, say that the electron has a charge of 1.602×10^{-19} C, then what is needed to explain the success of acting upon that is – wait for it – that the electron indeed has a charge of 1.602×10^{-19} C. Again there is no mention of truth anywhere. But this is science's own explanation of its own success, and there is none better.

If the conservative wanted anything else, this is probably something to diagnose and cure rather than to worry about. Perhaps he hankered after authority. He wanted an excuse from the heavy burden of judgment. He wanted the world to tell him what to think. He wanted there to be a Book of the World, a voice in things, which he had to obey. But there is no such voice, and could be no such voice (even the voice of God would be insufficient at this point, as Plato saw. For if God has not read the Book of the World, his voice will be arbitrary too: just another political pressure as it were). Young people become relativists when they realize that the voice of their parents and teachers is not the voice of the world.

The conservative hankering cannot be satisfied. But that is not a victory for the relativist. Remember the relativist's position is above the fray, taking the lordly, God's eye view of any debate. The relativist interjection "That's just your opinion" is not a move even *within* a moral debate. It's a move to close the debate. But some debates cannot be closed, since over some things we cannot agree to differ. Agreeing to differ is often a way of denying that something is a moral issue after all, but perhaps a matter of taste or lifestyle. If you think capital punishment is a good thing, and I think it is never under any circumstances permissible, then we differ, and we may have to sort out our differences. It is not a matter of taste or lifestyle. If we try to sort the difference out, the relativist interjection to each of us is useless. Of course we are putting forward our opinions, since that is what making an assertion is, unless of course we are lying or insincere. But we are putting them forward as opinions to be shared, or rejected. We are putting them into a public space, a space of discourse and reason. Their survival in that space is not a matter of their having a hotline to the Forms, or the Book of the World. Their survival is a matter of what to think: whether capital punishment is permissible, or capital punishment should not be allowed, or whatever the issue may be. The issue is the issue, not the metatheory.

Relativism attracts suspicion and hostility for a good reason. Suppose I voice an honest and heartfelt opinion about anything, from mathematics to aesthetics. The conversation stopping remark "that's just your opinion" is not only beside the point, but more importantly dehumanising. It signals that your words do not deserve to be taken seriously, but only taken as symptoms, like signs of a disease. Taking your words seriously would mean absorbing them into my own decision-making process, either finding them helpful or finding them to need rebuttal. They would become a factor in making up my own mind about whether, say, capital punishment is permissible or not. But if I look at you and see only the symptoms of a wishy-washy liberal ideology, or equally a harsh conservative vengeful one, then I am simply sidelining your words so far as the issue at hand goes.

This is why, for instance the "science wars" between honest-to-God working scientists, and allegedly debunking postmodernist historians and sociologists of science were so venomous. The scientist says, for instance, that the moon is about a quarter of a million miles from the earth. The historian or sociologist listens, but then goes into a story about how such a saying is the expression of an ideology or a perspective which itself arose for some identifiable sociological or historical reason: to assist the mercantile classes or the expansion of Europe, or whatever it may be. To the scientist this is insulting, since from his point of view the prior issue is how far away the moon is, and only then, and marginally, is he interested in the historical question of how people come to be convinced of it. His history of how it came to be believed starts with it being true: it is believed because that is how far away the moon is, and some very smart people were clever enough to cotton onto it, by finding ways of representing the distance in measurable terms.

By bracketing the question of truth, the historian or sociologist becomes exactly like someone who asks why I believe that there is cheese in the fridge, expecting the same explanation to be available, whether there is cheese there or not. Whereas two very different explanations are needed. There is a boring one for people who believe there is cheese there when, and only when, there is, along the lines of "they looked and saw it". There is a more oblique explanation for people who believe there is cheese there when there is none. This may be more or less startling depending on how explicable the mistake was: bad light, something that looked like cheese, or out-and-out hallucination. But any man with a white coat who resolutely brackets the question of truth, and then seeks to explain why I believe there is cheese in the fridge is treating me as a patient or potential lunatic from the word go. This is why he is dehumanising.

The relativist is in exactly the same position in ethics or aesthetics. He looks at us, and rather than hearing our words as they need to be heard, as laying down a position to be taken up and assessed, he hears us only making noises. We are, in Peter Strawson's tremendous phrase, there to be "managed or handled or cured or trained". No wonder, then that the relativist in turn excites such hostility. It is not only the conservative half of each of us who cannot stand this patronage. It is each of us in toto, agents attempting to reason our way through the practical problems with which life tries to trip us up.

Interestingly, and ironically in view of the supposed opposition between 'analytical' and 'continental' philosophy I believe that deflationism is where Nietzsche arrived towards the end of his life. Certainly, in many of his writings Nietzsche is on the side of scepticism or relativism, and he has been a kind of patron saint of postmodernist debunking, the great master of suspicion. Yet the persistent power of the ideal of truth, what he called the ascetic ideal, was one whose survival in the age of pragmatism and Darwinism puzzled him at the same time as it compelled his admiration. As he saw in his final work, *The Twilight of the Idols*, if we can get past old oppositions, then truth becomes, like God, the empty ghost of an idea. A substantial ideal of truth has absconded, even as an ideal, but as with God, this should not afflict us as a loss, but be celebrated as a gain, a liberation:

> 5. The 'true' world – an idea which is no longer good for anything, not even obligating – an idea which has become useless and superfluous – consequently a refuted idea: let us abolish it! (Bright day; breakfast; return of *bon sens* and cheerfulness; Plato's embarrassed blush; pandemonium of all free spirits.)
>
> 6. The true world – we have abolished. What world has remained? The apparent one perhaps? But no! *With the true world we have also abolished the apparent one.* (Noon; moment of the briefest shadow; end of the longest error; high point of humanity; INCIPIT ZARATHUSTRA).[4]

Like Nietzsche, I have tried to diagnose a shared illusion, common both to conservatives and their arch-enemies, the relativists. Where does that leave us? It may look as if it leaves us with all kinds of inquiry on all fours with each other, for deflationism is inconsistent with the belief I mentioned at the beginning, that there is a hierarchy of disciplines of which some have a more privileged relationship to truth than others. The notion of truth is too cheap for only one kind of discipline to purchase it. But although the notion is cheap, the truth about any particular matter may not be, so deflationism is consistent with a version of that hierarchy. There are disciplines, such as mathematics or science, within which convergence of opinion is a more striking phenomenon than in others. In disciplines of interpretation, such as sociology or history, and perhaps still more when questions of value dominate the stage, diversity of opinion seems more salient than convergence, and just as in classical times, this can engender pessimism and scepticism. In science an ideology, or narrative, of convergence is a good deal more credible than in interpretation, or ethics or aesthetics, for example. We should notice, however, that while this is true, it does nothing to encourage one corollary, which is too often drawn from it, which is that all we need to do is to apply something called the methods of science to matters of human nature, human history, human ethics or politics, and the same convergence will be forthcoming. For it is far more probable that science is convergent because it deals with things like atoms and proteins, which are less mutable and self-defining, than human beings. I suppose that a distinguished scientist pleased with his place at the top of the hierarchy might counsel that we stop doing serious history or political theory or ethics, but that would not itself be a piece of distinguished science, but a piece of very bad advice.

So we are left with what I have called the burden of judgement. Just as it took work and intelligence to determine a value for the charge on the electron, so it takes work and intelligence to think whether capital punishment should be allowed, or how to interpret a historical text or a piece of a constitution. I believe the retreating

---

[4]That is, Zarathustra arrives or begins his ascendancy.

seas of postmodernist controversy can also leave a valuable understanding of where we are. In the heat of the science wars, any attempt to introduce a sociological or historical dimension into thinking about science, its emergence in the West, or the social and economic structures that underpin its activities, was apt to be met with a blanket accusation of relativism. But now that the charge dissolves, we should be able to see that in themselves such explanations are neither sceptical nor debunking, nor even disrespectful to the astonishing achievements and successes of science. Let us then close by keeping in mind two thoughts:

The electron indeed has the same charge in London or Paris, Delhi or Beijing.
It took a fortunate confluence of political, social, economic and cultural forces to enable us to find it out.

Those two remarks are perfectly consistent, and each has its own importance. The second should remind us of the fragility of knowledge, for it takes a not dissimilar confluence of political, social, economic and cultural forces to enable new generations even to understand the achievements of the past. Hence, it reminds us of the burden that we academics bear as we try not only to increase those achievements but even to protect what we have got.

Sergio Romano:

Thank you very much, Professor Blackburn. Thanks to you we couldn't have a better beginning for our conference, and we have a great deal of food for thought.

Question:

You have mentioned capital punishment addressing the question whether it is licit or not. It is something we, as individuals, are never confronted with in our daily life as we never have to ask ourselves whether capital punishment is right or wrong. It is a matter for the state, not for the individual person. So the way you put the question about truth with reference to capital punishment is slightly misleading. I am surprised that you adopt the problem of the death penalty as the central problem to your explanation about truth. In other words, you introduced an element that does not concern an individual person but that concerns the state. Is it right or wrong for the state? It is not a personal issue.

Simon Blackburn:

There is a rather discreditable reason why I picked the example in my book on truth, which is that the book was to be sold in the United States.

I take it as important for each citizen, each member of a state, to have a view about these things. I agree, of course, that in Europe, the issue of the death penalty is dead, it's settled and finished. But worldwide, it is not. I wish it was. So it seems to me an important issue. But of course I could have taken a much more personal issue. I could have taken, for example, the issue of abortion, or issues which worry people at present, such as the propriety of immigration, things which people do talk about perhaps much more frequently than they talk about the death penalty. I wanted a straightforward moral example, but also an example on which, when it arises, people cannot agree to differ, because agreeing to differ must be, inevitably, coming down on one side or the other. You're either going to allow it or you're not. A state is either going to allow it or it's not, and you can't just shrug your shoulders and walk away. Of course it's a dead issue in Europe, I agree, but it's not a dead issue worldwide.

Question:

There's a convergence of scientific opinion that the moon is a quarter of a million miles away from the earth. Does that mean that fact is true?

Simon Blackburn:

Yes. In saying that, I'm just repeating that the moon is a quarter of a million miles from the earth, as far as I know. Approximately.

Question:

This is just a question about the deflationist or minimalist view that you defend. In a way, we in the religion section could be happy with that, given that in medieval philosophy, *verum* and *ens* are transcendentals and virtually equivalent, and what we're interested in is the reality. But I don't think you can do away with truth entirely, because truth is a property of beliefs and propositions, and we are interested in both thought and reality, not just reality – our thought about reality, our attempt to discover what reality is. When you were talking about the moon and the cheese in the fridge, you were using the word 'true' in the normal sense, despite what you said about deflationism.

Simon Blackburn:

The deflationist claims to give the normal sense. According to the deflationist, your understanding of the notion of truth simply consists in your disposition to accept any instance, or any – logicians would say any non-pathological instance, for reasons which needn't detain us – of the schema 'If *p* is true, then *p*'. Or '*p* is true if and only if *p*'. That's all there is to your understanding of truth. So it's not as if truth comes in as a second-order topic. That's the important point. However, there is something I didn't say today, and this is why I managed to write a book about the subject, which is that you may have to cast a beady eye on your own vocabulary, because the vocabulary with which you frame questions may be a vocabulary which is not adapted to coming to a judgment. So the burden of judgment cannot be discharged. Let us consider how mumbo-jumbo conquered the world. And if your head is full of mumbo-jumbo, you think you're expressing genuine assertions, genuine candidates for truth, when you're not. And I would say most of theology has that defect, though others may not.

Session I: Philosophy

16 May 2008 – 1:30 pm

Chair:
Salvatore Veca

Speaker:
Emanuele Severino *Truth, Contradiction, Nihilism*

Panel:
Pascal Engel; Karlheinz Stierle;
Marcelo Sánchez Sorondo

Salvatore Veca:

Professor Blackburn's lecture has been an excellent onset of our conference. Now we will focus on truth in a contribution by Emanuele Severino followed by contributions from panellists. I would like to give the floor to Professor Emanuele Severino, professor at the University of Venice and San Raffaele University of Milan.

Truth, Contradiction, Nihilism

Emanuele Severino:

I

In this day and age, an increasingly dominant conviction has been gaining ground that truth – whatever form of truth – has a historical and pragmatic nature. Truth is not above and beyond time and history, rather it is a certain temporary state of knowledge, which nevertheless endures and is affirmed insofar as it can achieve certain goals.

This view is the negation of the characteristic of incontrovertibility, universality and necessity that, starting with Greek philosophy, throughout Western tradition, and not only in the field of philosophical thought, has been assigned to truth. Today, true knowledge is that which in certain established spatial-temporal circumstances

Balzan (ed.), *Truth in Science, the Humanities and Religion: Balzan Symposium 2008*,
DOI 10.1007/978-1-4020-9896-3_2, © Springer Science+Business Media B.V. 2010

succeeds in predicting and transforming the world more than other forms of knowledge. Truth is power, action, winning praxis, and in our age, winning power is held to be technical knowledge guided by modern science. It is thought that the growth of scientific power is determined by the progressive modification and substitution of scientific theories, all of which tend – even those which used to be considered incontrovertible truths – to acquire the characteristics of statistical-probabilistic laws, or of historical and pragmatic truths. Mathematics, too, recognizes that its principles are postulates, hypotheses that do not claim to have absolute value.

On the other hand, the thesis that every truth has a historical-pragmatic nature cannot get round the so-called 'argument against the sceptic', according to which this thesis, presenting itself as incontrovertible truth, refutes itself. In order to avoid such a consequence, that is to say, such a contradiction, the thesis of a historical-pragmatic nature for truth ends up presenting itself as historical-pragmatic truth, controvertible. Much earlier than when this view was set forth by philosophers in the analytical tradition such as Richard Rorty, it had been put forward in Italy by philosophers like Ugo Spirito, a pupil of Giovanni Gentile. It should be noted, however, that it is precisely because the thesis of the controvertibility of every truth was not abandoned that it was proposed to give it a non-contradictory form, recognising the controvertible, historical-pragmatic nature of this same thesis. Whereby it is, however, clear that in this apparently radical form of repudiating incontrovertibility, something of incontrovertible truth can be recognised in the argument against the sceptic; and since this argument shows the presence of a contradiction in the absolute thesis of the controvertible nature of every truth, in that apparently radical form of negation of the incontrovertible one can also recognise an element of incontrovertibility in the principle of non-contradiction.

By these considerations it is intended to say that the authentically more radical form – even if tendentially unobserved – of the philosophy of our times is something else: it is neither scepticism nor that apparently radical form of negating the truth that recognises its own controvertibility, historicity, pragmaticity. In its authentically more radical form, the philosophy of our times does not negate every incontrovertible truth, but negates every incontrovertible and immutable truth that claims to place itself above the one fundamental incontrovertible truth that is permanent as long as something exists, the truth that consists of the thesis that every truth that is different from this thesis is overwhelmed by time, by history, by the becoming of the world.

At this point, it is a question of seeing why this is the only fundamental incontrovertible truth. Since as it is not enough to assert, as for the most part happens today, that there does not exist any meta-historical, metaphysical, absolute, definitive truth, that there does not exist any immutable and necessary Being, any Foundation, any absolute Centre, any absolute Meaning of the world. Today for the most part all of this is taken for granted. But in its most profound essence, the philosophy of our time deliberately shows the necessity of all of this, and it shows it on the basis of the belief that the existence of becoming and of the historicity of every thing and every truth is the supreme incontrovertible truth, which is itself destined to annihilate itself when there will no longer be any consciousness of the world.

However, it is not on the necessity of the philosophy of our times that we intend to dwell, but upon the dimension that is common to both the traditional conception of truth and to the destruction of this conception (that is, to the destruction worked upon it, in fact, by the essence of the philosophical thought of our times). We intend to show, albeit from a distance, a sense of truth that does not belong to the history of the West (much less of the East), but that has always appeared in the innermost recess of each of us.

II

The Greek word that we translate with 'truth' is *alétheia*, which, to the letter, means 'not hiding itself', and therefore, the displaying, the revealing, the appearing of things. In Greek thought, however, truth is not only *alétheia* (unlike what Heidegger maintained): truth is that appearing in which that which appears is the incontrovertible, or that which, says Aristotle, "cannot be in any other way than as it is and is present" (*mè endéchetai àllos échein*). This 'being' in the absolute is expressed in Greek philosophy with the word *epistéme*, where **steme* (from the Indo-European root **sta*) indicates in fact the 'being' of what is and what imposes itself 'on' (*epi*) every force that wants to negate it, shake it, strike it down. That which cannot be in any other way is incontrovertible, it is the way things are. Exposing oneself to the possibility of being otherwise, by falling, is the trembling of thought. That is precisely why Parmenides says that the "heart" of truth is "intrepid" (*atremès êtor*). The heart of mortals, instead, trembles in the face of suffering and death. Plato and Aristotle call this tremor *tháuma*, and they see that philosophy is born from *tháuma*, that is from "anguished wonder": in order to be certain of salvation, one cannot be satisfied with myth; it is necessary that true salvation come forth.

For some time now, philosophical thought has realized that the traditional definition of truth as *adaequatio intellectus et rei* ("correspondence of the mind and reality", "correspondence of the intellect to reality") has a subordinate nature. In fact, in order to ascertain whether the intellect is commensurate with reality, it is necessary that reality manifests itself, and that it should appear not as the content of an opinion or of a belief, but in its not being able to be otherwise, that is to say, in its incontrovertibility. Idealism, neo-idealism, phenomenology are aware of the derivational or, in fact, altering nature of the concept of *adaequatio*. Even the philosophers of the so-called 'linguistic turn', although they have a hard time recognising this, have been going in the same direction. Greek thought, too, in which the form of the conception of the truth as 'conformity' moves back beyond this definition and, as has been noted, conceives of truth as the union of *alétheia* and *epistéme*, where the things of the world that first of all manifest themselves, are incontrovertibly mutable, move from non-being to being and vice versa, and in this sense can stand otherwise than they do, nevertheless so long as they stand in a certain way, it is impossible that they should stand otherwise and in this sense, they too are incontrovertible. Moreover, *epistéme* holds that starting with that which manifests itself, it is necessary to arrive at the affirmation of the divine, immutable Being, distinct from the world of becoming and, within the world of becoming, from the affirmation of the existence of structures and immutable forms that in some way reflect the immutability of the divine: the so-called 'laws of nature', like 'natural' law and 'natural' morality. In the *epistéme*, the standing of God and of God's reflection in the world is different from the stability and permanence of myth, precisely because epistemic standing does not appear within a belief, or a will to let something stand. A supreme, divine Being is authentically immutable only insofar as its immutable existence is affirmed by incontrovertible knowledge.

What this means is that the sense of the incontrovertible brought to light by Greek thought, is not to be confused with a law whose inviolability is, on the one hand, the simple will that such a law should be inviolable and, on the other, the punishment that will be meted out to those who break that law. These are the laws that regulate the life that is guided by myth or everyday life, and that, in the eyes of philosophy, begin to appear as simple social habits, forms of desire, of violence, of belief and thus as something controvertible. Undoubtedly, for *epistéme*, whoever thinks or acts against truth (that is, against the incontrovertible) goes towards his/her own ruin, but the incontrovertibility of truth does not consist in being able to think or act in a certain way without being punished or destroyed. Hence incontrovertibility does not consist in the advantages that it procures (even if for *epistéme* only those who live in truth can be happy). Nor does it consist of an impulse or a biological constraint that only exists as long as humankind does, or until a certain kind of human being exists, and that may no longer exist should humankind or that certain type of human being no longer exist. The incontrovertibility of truth consists in impossibility: the impossibility, to take up Aristotle's expression again, that that which is manifest be otherwise than as it reveals itself to be.

The root of this impossibility has been shown since the beginning of Greek thought: in the language of Parmenides this root is the identification of 'the being' and of 'the not being'. In Book IV of Aristotle's *Metaphysics*, where the 'principle' of the *epistéme* of truth is shown, the root of the impossibility of being in any other way is the negation of that *bebaiotáte arché* (*principium firmissimum*) that will henceforth be called the "principle of non-contradiction". The Greek word *básis*, upon which the superlative *bebaiotáte* (*firmissimum*) is constructed, indicates the standing of the incontrovertible. In its original form, the impossible is the negation of the principle of non-contradiction. In fact, this principle states that it is not possible to be in a way that is otherwise than what one is. That is precisely why being is a standing.

However, more precisely, in Greek thought the root of the impossible – the root of that of which *epistéme* is the negation – is the synthesis of the negation of the principle of non-contradiction and the negation of the beings that appear and that make up, first of all, the world of becoming. The root, the foundation of *epistéme*, is the synthesis of the principle of non-contradiction and of appearing (that is, of *alétheia*) of the beings of the world, that is, of the beings that temporarily leave their state of non-being and return thither.

From the root of *epistéme*, the trunk of *epistéme* develops, which culminates in metaphysics, that is, in the affirmation of the immutable and divine Being.

And the root of *epistéme* also becomes, more or less consciously and explicitly, as well as more progressively, the premise of every thought and every action of Western man. It also becomes the premise of scientific knowledge. What in fact can any science stem from if not from the appearing of the world (that is, from *alétheia*)? In this appearing of the world (in accordance with the principle of non-contradiction) every being is identical to itself, that is, to every being that appears may not pertain and at the same time may not fail to pertain those properties that at a certain moment in time pertain to it (including such properties as those indicated by Heisenberg's "principle of indetermination").

Of course, not even *epistéme* comes out of a vacuum, but in turn presupposes the appearing of mutable things in the world; but it is *epistéme* and only *epistéme* – that is, only philosophy – that may conjure up the meaning of truth, that is, of the incontrovertible. Philosophy may well acknowledge the fact that it has been preceded by the appearing of the world. But, in acknowledging this, it not only fundamentally transforms the world, but also catches sight of the non–truth (the controvertibility) of the (prevalently mythical) pre-philosophical interpretations of the world. Husserl's concept of the "world of life" moves in this direction.

Yet Husserl, and he was not alone in this, did not realize that the world that appeared before philosophy is transformed by it in a radically more profound way by the birth of philosophy. It is to this more profound transformation that we must now turn.

III

Especially in the philosophical and scientific thought of the past two centuries – but also in religious and artistic thought – there are many ways of negating the incontrovertibility of the *principium firmissimum* (that is, of the principle of non-contradiction – of identity – of the excluded third). Yet the sense that Greek philosophy gives to this principle remains the foundation of the entire development of the West, and thus also of the culture and civilisation of our time, which for that matter arrives at the most radical destruction of what we previously called "the trunk of *epistéme*" and the "metaphysical-theological culmination of that trunk". It is a question of understanding that the critics of the principle of non-contradiction assume this principle as the basis of their discourse precisely because they, too, affirm the existence of the becoming of the world; they assume that principle as their basis even if they do not often realize it, and despite their conviction (think, for example, of Nietzsche) that this principle is incompatible with the becoming of beings.

The destruction of the trunk of *epistéme* is the negation of the possibility of any immutable and eternal Being: it is the 'death of God'. After such destruction, it is necessary to affirm that every being exists in time, is more or less ephemeral, is corruptible; and this statement is at the same time an affirmation that every truth is historical, temporal, temporary, mutable, becoming. Nevertheless, for the destroyers of an immutable Being, the existence of becoming, of time, of history, is not a simple opinion and controvertible belief: with different degrees of awareness, the existence of becoming is also, for these destroyers (as for the entire Western World), incontrovertible truth, indeed, it is the only incontrovertible truth. The only incontrovertible truth is the overwhelming – the becoming, in fact – of every truth.

And the incontrovertibility of the existence of becoming maintains precisely that sense that pertains to incontrovertibility insofar as it constitutes itself as a root, principle, foundation of *epistéme*. So much so that even for the thought that moves the trunk of *epistéme* towards decline, the existence of becoming is an incontrovertible truth because, on one hand, becoming appears, is manifest (it is the original content of *alétheia*) and because, on the other hand, the beings that become exist in accordance with the principle of non-contradiction. In fact, there is becoming only insofar as the beings that exist 'beforehand' are not the beings that exist 'afterwards', and therefore only insofar as the dimension of the 'beforehand' and every element of this dimension are themselves and not otherwise than themselves. The transformation of beings is possible only if the beginning of the transformation is not identical to its completion. But throughout the entire history of Western thought, the principle of non-contradiction affirms, in fact, that each being is itself and is nothing other than itself as long as it is what it is. For example, it is necessary that a surface be white and not black as long as it is white, that is to say, for as long a time as when it is white; but then it can be transformed, and at a different time it may be black. It is necessary that man be man as long as he exists; but then he dies, and at a different time, he is not-man because he has turned into ashes, nil, nothingness.

Since the beginning – thus even before philosophy – it has been thought that things transform themselves, become something other than what they are. Adam believed he could become God. But philosophy, in being born, thought for the first time that things are 'beings' and that a being is not nothingness. For the first time philosophy thought about the being in relation to the most radical meaning of nothingness: at the centre of the principle of non-contradiction lies the opposition between being and nothingness. This means that with the coming of philosophy, things start becoming something other than themselves in an unprecedented way: when a white surface becomes black, the white surface turns into nothingness, that is to say, its specific being this white surface turns into nothingness: this specificity turns into nothingness even if something of this white surface does not turn into nothingness – for example its being a surface does not turn into nothingness when the surface becomes black. When a man dies, his specific being a man turns into nothingness even if his corpse is not nothing and even if something of him should continue to exist – and even if his corpse should rise again; since the risen body and in general the 'promised land' mentioned in the *Apocalypse* of St John the Divine are not this body and this land, which are also definitively destined to turn into nothingness in the eyes of so-called Christian 'optimism'.

The principle of non-contradiction, which has dominated the entire history of the West, thus affirms that it is necessary that beings be beings, as long as they are, but that it is not necessary that beings be; which means that it is necessary for beings to have been nothing before being, and that they return to being nothingness, after having been. Thus the principle of non-contradiction is, at the same time, the affirmation that the beings that manifest themselves become, that is, leave nothingness and return thither; and in its turn, becoming, as it is understood by Western thought, is at the same time the affirmation of the principle of non-contradiction, that is the principle whereby it is impossible that, when beings are, that they should be nothingness, but whereby it is necessary that at another time, beings should have been and should return to being nothingness. In the theological-metaphysical form of *epistéme*, only the divine is "always safe" from nothingness (as Aristotle says); in the destruction of such a form, that is, in thinking the 'death of God', every being comes out of nothingness and is destined to return there. But the fact that beings, insofar as they are beings, leave nothingness and return thither is a truth shared both by the Western tradition and by the destruction of that tradition.

IV

Yet, beyond the meaning that the West has assigned to truth, that is, to the incontrovertible, truth and the incontrovertible show an essentially different sense – so different as to force us to state that the history of the West is the history of the most profound alienation of truth, that is, the history of nihilism – and, this, once again, in an essentially different sense from what this statement means in the philosophy of Nietzsche and of Heidegger. This sense of truth, that lies beyond the West, and indeed even beyond the whole history of mortals on Earth, is not the theoretical product of one individual or of one people, nor does it have anything to do with any form whatever of the divine: this sense has always appeared and always will in the innermost recess of us all, even if language does not turn to it and does not bear witness to it. The alienation of truth is not simply a different landscape from that which truth consists in, but it is the appearing of truth within an atmosphere that causes the distortion of the face of truth, making it unrecognisable. Let us try to clear up, at least from a distance, the meaning of these statements, which may certainly sound arbitrary and out of joint.

Even though one can hardly become aware of it explicitly, in every field of knowledge and of action the opinion has become prevalent today that every being at a certain moment ceases to be, that is, – as stated above – that every being becomes other than what it is, and becomes other in a such way that its specificity turns into nothingness. This is also the case even if some traits of this specificity remain. For example, when death annihilates the specificity of a human, the corpse is left over, or, to speak in terms of physics, the quantity of energy that constitutes this human is not annihilated, but remains. When instead a human begins to be, nothingness is turned into that being that is the specificity of this human, even if the masculine and feminine gametes from which he or she will develop already exist, or, in other words, even if the quantity of energy that will make up this human already precedes the moment when this human begins to be. Nothingness – it is thought – turns into being and being turns into nothingness.

But – and here we come to the crunch – when a being, for example a human, is annihilated, that is, turns into nothingness, the result of this becoming is not simply nothingness, but is the nothingness – of – this human. This statement is different from the other statement expressed above, that as a result of annihilation the corpse is nonetheless left over. This new statement says that the result of becoming, that is to say of annihilation, has come about because the content of the result stands in relation to that from which such content results, and that is, because, as was said in the example, the result is not simply nothingness, but the nothingness – of this human. However, the expression "the nothingness – of this human" means: "This human is nothing". As a result of becoming other, this human is nothing. In general, as a result of becoming in which the being is turned into nothingness the being is nothingness; or, also, the being is not-being; or, also, the being is not being. Similarly, in becoming being by coming out of nothingness, the outcome of this becoming is not simply a being (for example, this human) rather it is nothingness that has turned into this human; so that as the result of this becoming nothingness is being (it being understood that, here too, the statement that nothingness becomes being is different, but not incompatible with the statement that some traits of the being that begins to be precede this beginning). The most radical alienation of truth, the utmost folly, is the conviction that being, insofar as it is being, is nothingness, and to live and build civilisations according to such a conviction.

What emerges therefore from these considerations, is that the principle of non-contradiction is contradictory, and that is to say that becoming is contradictory, in that becoming is to become other, and with the arrival of philosophy, it is to become other absolutely, that is, being turns into nothingness, and nothingness turns into being. The principle of non-contradiction is contradictory, and that is to say becoming is contradictory. And vice versa.

We have seen, in fact, that the principle of non-contradiction is the very principle of becoming and that this principle is the fundamental sense of truth throughout the history of the West, which has now become the history of the planet. This principle may very well state that it is necessary that being be being and that it not be non-being, but it also states that that is necessary when, for as long a time as when being is, and thus it is necessary that during the time when being is not, being be nothingness, and that, during the time when being is, nothingness be being. In spite of every attempt to demonstrate the contrary, time, that is, becoming, belongs to the essence of the principle of non-contradiction. And since – this too was brought out above – there is becoming only in as much as what is beforehand cannot be that which is afterwards, and since the nothingness that is beforehand cannot be the being that is afterwards, and vice versa, it turns out that for this side, too, becoming belongs to the essence of the principle of non-contradiction.

But now it has come to light that the principle of non-contradiction, and that is, becoming, is contradictory, that which in the eyes of Western thought is the supreme and original truth is contradictory. The synthesis between the principle of non-contradiction and the appearing of the becoming of beings is for the West the essence of truth, incontrovertible; but now it has come to light that which for the West is incontrovertible is absolutely contradictory, impossible, the alienation of truth, the very essence of nihilism. Nihilism means precisely thinking that being is nothingness; and the conviction that being is nothingness is the hidden foundation of the belief in which the West wants to proclaim, while hiding its own foundation, that being, inasmuch as it is being, is not nothingness. Nihilism is also the very same criticism of nihilism by all of the inhabitants of the West, who still firmly maintain that the becoming of beings is their coming out of nothingness and their returning thither. Nihilism is not only the criticism of nihilism by Nietzsche and Heidegger, but also by Christianity and the other religious forms that have inscribed their contents in the Greek sense of being and of the becoming of being.

The authentic sense, that is to say, non-nihilism, of being oneself and of not being other than oneself, of every being – the authentic sense of truth – cannot therefore be the sense that the West has assigned to being oneself and to not being other than oneself. We shall call this authentic sense of being oneself and not being other than oneself the "destiny of truth".

V

Western thought realizes that every form of knowledge and action (religious, technical, metaphysical-theological, artistic, scientific) must be founded on the manifestation of the world, but ever since its birth, philosophy has enfolded this manifestation in the most radical form of becoming other, that is to say in the most radical form of the alienation of truth, and it is this alienation that in its turn has enfolded wider and wider dimensions of the becoming of the world, bringing to light a course that constitutes the nature of the history of the West which, for that matter, is extending its reach to the entire Planet.

The civilisation of technology, to whose essence of necessity belongs the philosophical thought of our time, is the most rigorous form of the most radical form of the alienation of truth. To say that such alienation lies at the heart of science obviously has nothing to do with the naive pretension of telling science in which way it ought to proceed; but to say this means to show that every power, and hence the supreme power of science and technology is possible only when founded on the alienation of truth. Only if one believes that beings are in and of themselves a coming out of and a returning to nothingness can one propose to methodically plan their production and their destruction. For some time already, science has abandoned its will to be the incontrovertible truth of *epistéme*. So much so that science realises that such will reduces and weakens prediction and power, that is, it impedes that continuous modification of the hypothetical, temporary, falsifiable theoretical apparatus which has made possible the growth of prediction and power. In this sense, science may have absolutely no objections to the thesis that the power of science and technology is not simply due to the controvertible and hypothetical nature of scientific assertions, but, in fact, to the alienation of truth, that is, to the alienation of what lies at the root of these assertions, too.

On the other hand, the existence of power – of every form of power and hence of the power of technology led by modern science – is also not incontrovertible truth in the sense the West has attributed to that which is incontrovertible. In fact, it has always been thought that no power exists but for the 'public recognition' of power; but that something like 'public recognition' should exist is not self-evident simply by its manifesting itself and being therefore indisputable: rather it is an interpretation, that is, ultimately, a belief, the will that such recognition should exist in the world. That power should exist is hence a belief. And this belief (like any other belief) believes itself to be incontrovertible truth. Only because one believes in the existence of becoming other can one believe in the existence of power and intend to be powerful.

However, not only is the existence of power not incontrovertible truth, but rather is something the appearing of which is authentically impossible, that is, it is something the appearing of which would be the negation of any authentic sense of truth, a negation of what we called above the "destiny of truth". In fact, if the alienation of truth is the becoming other of beings, non-alienation, that is, the destiny of truth, is the non-becoming other of beings, and hence their not coming out of nothingness and their not returning thither. The destiny of truth is the eternity of every being, that is, the eternity of every state of the world and of the soul, of every situation, relation, instant, it is the eternity of all that is manifest or hidden. The authentic being oneself of every being, its authentic not being other and not becoming other is its eternity.

The Gods are jealous: they keep eternity for themselves and abandon the world to nothingness. In view of the destiny of truth, instead, even the most humble and reclusive of beings is eternal. But this does not mean that the variation of the world is only a delusional opinion, as held by Parmenides and Eastern thought: it means that coming out of nothingness and returning thither, by the world's beings, is not and cannot be something that appears – and hence the variation of the world is the coming of the eternals in the eternal circle of appearing.

The sun crosses the vault of the sky. But when the sun goes down, the vault of the sky cannot show what fate has befallen the sun once it has gone down. Every being is a sun that crosses the 'vault' of appearing. When beings disappear in accordance with certain ways that mortals have been fearing for some time now and that are the various forms of death, mortals believe that beings have gone to nothingness. But this belief is a theory, or, again, an interpretation – the interpretation that lies at the heart of every other interpretation. Horrendous and anguishing though the 'sunset' of beings may be, that is, their leaving the 'vault' of appearing, this 'vault' does not show and cannot show what fate befell that which left it and no longer appears. This fate is shown by the destiny of truth, according to which every being is eternal and hence it is impossible that its disappearing should mean going into nothingness.

Nor can the power of drawing out beings from within nothingness and pushing them back into nothingness therefore be something that appears. Power is only one content of the interpretation, and nevertheless even the interpretation is a being and hence eternal. And the interpretation in which the clash between the different forms of power appear is also eternal. Nihilism is also eternal. And at this point it is a question of understanding in what sense the eternals that constitute the content of the alienation of truth are traversed by the eternals of the destiny of truth.

But even looking back, towards the themes that we have touched upon, how far away they still are, the indications that we have been able to call up, from the concrete sense of the destiny of truth! Destiny is, on one hand, the appearing of the eternity of every being, implied by not being and not becoming other than oneself and, on the other hand, it is the appearing of the coming of the eternals into that being that is the eternal circle of the appearing of beings. But that the being should not be other that oneself and should not become other than oneself is not a postulate, a dogma, a belief, a hypothesis. It is necessity ('destiny', in fact). But it is necessity only insofar as the impossibility appears of negating the being oneself and the appearing of beings, and only insofar as the authentic sense of this impossibility appears. It is only starting from this point, that is, from the point at which we have arrived, that the destiny of truth can start to show its own concrete sense.

Panel Discussion

Salvatore Veca:

Let me just thank Emanuele Severino for his contribution and hand over to Pascal Engel, professor of the history of modern and contemporary philosophy at the University of Geneva.

Pascal Engel:

I would like to comment on the words of Professor Severino and I also have some comments about Professor Blackburn's speech.

Professor Severino's contribution is an in-depth analysis, and it's extremely difficult for me to ask direct questions or comment on this contribution. There are two topics that I would like to address here: the first is the question how much does the truth depend on mind, and the second is about the idea of truth as identity.

Heidegger states that before Newton's laws were revealed, they were not 'true', but that doesn't mean that they were false. They were not revealed, and so these theories were neither false nor true, but this does not mean that the revealed being was not a being before being revealed. This stand depends on the following mental approach. First: only beliefs and assertions can be true; second: beliefs and assertions depend on the existence of thinking minds. Hence, in a word, without thinking minds, there is nothing which can be actually true. So the very first question that I would like to ask Professor Severino is the following: would you accept the theory of identity and truth of the mind and the theory of truth and the justification of this truth?

The historical-pragmatic conception and the idea of incontrovertibility of truth seem to be associated with a thinking mind which is able to perceive this truth. So when we say that something is a belief or an assertion and that this is true using a certain word, we are not just talking about the truth of the belief, of the assertion. We are talking about this word, something which is there, which is true in that word. This conception requires objectiveness in each assertion.

The second question I would like to ask Professor Severino is the following: you have been talking about an implicit conception of truth. What about the Parmenidean conception of truth? Is there an identity between thought and being?

Emanuele Severino:

It's extremely problematic to state that. If you say that truth is a thought, then there is a certain form of idealism. If we state that truth is the word, then the word is a fact, and I think it's extremely problematic to state that.

Pascal Engel:

I found this talk extremely congenial to our issues today, but one question which I would like to pose is about Ramsey's ladder. Remember, Ramsey's ladder is where second-order questions about truth, as he calls them, are absolutely useless in front of first-order questions, which have to do with the way we judge, the way we believe certain things. And for that the only definition of truth that we need is the famous Tarski's schema of Convention T.

Now, Simon Blackburn seems to be against those relativists who laugh when the Catholic priests say, no this is the truth, this is the true God. But at the same time, he also has some sympathies with deflationism. The deflationist is the theoretician who believes that truth is nothing but a little tap, or a little compliment, as Rorty says, that we pay to our assertions. Blackburn has some sympathy for this kind of view, because he doesn't want to accept either the reductionist or the constructivist view according to which truth is just the expression of our feelings. At the same time he doesn't want to accept the realist view, according to which there is something, which has a property, and that this property is truth.

I have with Simon Blackburn's position more or less the same problem which I have with Professor Severino's, as far as I could understand it. It is this: it seems to me that he is quite right to say that when we want to answer the relativists or the deflationists, it is very difficult for us to bang on the table and say "this is true", because it just repeats the initial assertion. I would like to ask him whether he is actually able to recognize our commitments when we say that something is true. In that respect, I would like to defend this hypothetical Catholic priest who says "this is the truth". Why should we say such a thing? It seems to me that when we say that something is true, we incur certain commitments. In the first place, we commit ourselves to the fact that if we assert that *p*, we accept at least tacitly to maintain our assertion – against opponents, for instance. We commit ourselves to the fact that our statements are objective. We also commit ourselves to a distinction between truth and justification, between the fact that a certain statement is true and the fact that it is justified. And we also commit ourselves to the fact that we ought to believe that *p* if we assert that *p*. Simon Blackburn points out that this will not impress the deflationists, because he will be able to state that these normative commitments are just another banging on the table, by saying "this is true", this is normatively true, this normatively, normatively, normatively true… and then we ascend Ramsey's ladder once more. But it is just not correct to say that when someone says that there is a norm, or that there is a value of truth, this just registers the way we judge. It's not correct to state that when we say, for instance, that we ought to believe that *p*, if and only *p*, this registers a value that has to do with the fact that we praise truth or that it has an instrumental value. I would say that it registers something that has intrinsic value.

Bernard Williams, in his last book on truth and truthfulness, has used an argument which he also uses for utilitarianism. If everybody believed that we just incur some instrumental commitments when we speak about truth, then nobody would believe it, and it would be self-defeating. Blackburn has identified what we might call the "deflationist gambit", but it's not correct to say that when we just list a number of sentences or assertions of the form 'you ought to believe that *p* if and only *p*' – where *p* might be any sort of proposition about the moon, about values, about whatever – that it just amounts to repeating each sentence individually. We do register something like a property, there is something which is common between those assertions, and this thing which is common is not reducible to a sort of schema which you can just deflate. So, I would say that it's not true that realism about truth is a kind of theological view. We can afford to keep some of the properties of the theological view in a laicist context.

Salvatore Veca:

Professor Karlheinz Stierle, Member of the Balzan Prize Committee and Professor Emeritus at Konstanz University.

Karlheinz Stierle:

I am not a philosopher, but I am replacing a philosopher, Professor Wolfgang Künne, who unfortunately fell sick at the last moment. He would have been the great German specialist about all questions pertaining to the history of truth that one could ask.

My subject is literature which raises philosophical questions all the time. And one of these questions is whether philosophy might not be part of literature. Reading Mr. Blackburn's book, *Truth*, and listening to his lecture this morning, I admired the elegance with which he steers his ship called 'Truth' between the Scylla of philosophical skepticism and the Charybdis of philosophical absolutism, not listening to the sirens. One of the essential results of this thoughtful book I would like to retain is that there is not one over-arching truth, but there are different domains of truth with their own rules and their own languages. Each domain has its own 'norms of right procedure', and I quote Blackburn: "These govern whether you have done your stuff properly – taken the right observations, made the right inferences, hedged in the right places, weighted the evidence carefully and, in short, made yourself immune to procedural criticism."

I think this way of differing between domains is a form of pragmatic truth. It is useless, for instance, measuring the distance between London and New York in millimetres. We could perhaps be precise, but this precision would be of no use at all. So there are kinds of truth, or resemblances of truth which just don't have any worth. But on the other hand, with regard to domains you always have to pay a price to truth. If you refer only to domains truth cannot just be added to one domain after the other. In his book and also this morning, Professor Blackburn has given, if I understand him correctly, a kind of map of these domains. But I wonder whether this map is complete, or whether it has white spots which might be discussed.

One of these white spots, it seems to me, is what in shorthand terminology might be termed 'continental philosophy'. On reading Blackburn my subjective impression was that there is a continent of bright truths, and then a continent of shadowy truths, getting darker and darker while the distance from this bright truth is growing, and ending in the dark continent of nonsense – or of mumbo-jumbo.

Stressing the need for truth and referring to different domains, I would like to make a very simple distinction between study and science. *Studium*, in the Middle Ages, mainly refers to texts, whereas science, from the beginning of modern times, mainly refers to natural objects. Study, then, is a domain of relevance, especially relevance bound in texts. Cultural objects are not given, they have to be constituted, which creates quite a lot of problems regarding truth. Study, it seems to me, is firmly founded in language. You can never get rid of it. Science can try to free itself of a specific language, but perhaps only in order to become subject to another language. This might be a kind of Vico's paradox of *verum factum*. Truth, on the one hand, is possible only in the domain of man's own productions or fictions, and the opposite would be that truth is only possible in the domain of natural phenomena, whereas the *verum factum* is just illusion and fiction.

One could refer here to a very interesting essay by Paul Valéry, where he invents a sort of *objet ambigu* that Socrates might have found on the beach, not knowing whether this object is natural or cultural. Proceeding from the principal presupposition one gets into very different and distant domains. It seems to me that these domains very much differ from another.

There is a profound relation between truth and language. Truth occurs in propositions. However, a text about truth, or a true text, is never a sum of propositions – it is truth at work, or work on truth, and cannot be just divided into true or wrong sentences. Propositions do not exist beyond language. Every language has a different shape, and so has the word truth in every language. If truth is necessarily bound to language, this means that it can never be pure in an absolute sense, which also means that it can never be freed of connotations. *Wahrheit* evokes a different background of cultural memory than *alétheia*, *veritas*, *vérité*, *verità* or truth. Can we get rid of this burden by just declaring it to be nonsense? That seems to me very difficult.

I will give one striking example with regard to connotation: the designation of a part of Germany that once called itself DDR. In West Germany, you had no choice. With each designation, you indicated at the same time your position: if you said

"DDR", you said "I acknowledge this and I am a communist"; if you said the "so-called DDR", you said "this is not a state at all"; if you said the "Soviet zone", you said "this state is under occupation", and if you said "Middle Germany", you said "I do not recognize the Polish border"; if you said "East Germany", you said "I recognize the border, but I do not recognize this as a state". You had no choice to describe this object in an objective way, getting rid of your connotations.

A special problem in this respect seems to me to be fiction. Fiction, as we know, is contrary to truth. But lie and error are also contrary to truth. Where then is the difference? Do lie and error refer to truth in the same way as fiction? Fiction is chameleon-like in its attitude to truth. When in the first draft of the French Constitution we read the sentence "*Tout Français est soldat*" – 'every Frenchman is a soldier' – is this fiction or truth? Declarative sentences often are fictions before coming true. "I declare you husband and wife" – is that true, and who has the authority to constitute this truth? Is poetry then a kind of fiction, or a kind of truth?

In his essay *Y-a-t-il une verité poetique?* – Is there a poetical truth? – the Balzan laureate Yves Bonnefoy, poet, literary critic and philosopher of poetical consciousness, strongly insisted that poetry is the only language that is language in the full and original sense. In poetry the perception of the world is not alienated in the abstract, world-less language of the concept. He comes close to Hölderlin, who has said poetically that man inhabits this earth. Is this a true proposition, or is it a poetical one? When Boethius in *The Consolation of Philosophy* wants to talk about real truth, philosophy begins in poetic language, which is quite interesting.

Salvatore Veca:

We should now hand over to Marcelo Sánchez Sorondo, professor at the Libera Università Maria Ss. Assunta of Rome, who is going to conclude the session.

Marcelo Sánchez Sorondo:

I would like to give a parallel contribution to Professor Severino's about the status of the human being in the age of science. Let me just start with Kant. What is man? That's Kant's question: a question about man and the truth of man.

At the end of his laborious journey in the labyrinth of the transcendental 'I', his *Critique of Pure Reason*, Kant confessed, with the frankness of an ancient Stoic, to a deep feeling of expectation:

> The whole interest of reason (both speculative and practical) is centred in the three following questions: 1. What can I know? 2. What should I do? 3. What may I hope?[1]

[1] Kant, *Kritik der reinen Vernunft*, A 805; B 832; *The Critique of Pure Reason*. English translation (with my adjustments): cf. http://etext.library.adelaide.edu.au/k/kant/immanuel/k16p/k16p95.html.

And taking up these questions again in his later *Logik*, he added as a fourth question: "what is man?" In ultimate terms, he added, and his clarification was new and of an essential value,

> all of these answers could be attributed to anthropology because the first three questions are related to the last, namely: what is man?[2]

These questions made anthropology the privileged place for the search for truth and made Kant one of the most radical and brilliant thinkers and at the same time one of the most problematic. The laborious analyses of *Kritik der reinen Vernunft*, which Kant wanted to categorise as 'metaphysics', led to him being lost in a labyrinth which ends with more questions than answers. However, he does not surrender but appeals – probably under the influence of Hume's notion of 'feeling' which gave vitality and value to 'impressions' – to profound conviction and experience. As he manages to state in the famous text of *Kritik der praktischen Vernunft*:

> two things fill my spirit with an ever new and increasing admiration and veneration, the more my reflection increases: the starry sky above me and the moral law inside me.[3]

He has no uncertainty or doubt: "I see them both before me and I connect them immediately with consciousness of my existence". Two fundamental experiences and thus two paths of openness to the Infinite: each one with the same anthropological point of departure but with different directions that have become today, with the development of contemporary physics, astrophysics, biology and neurology, of particular significance and contemporary relevance. Kant explained all of this after what he himself called the revolution in physics of Copernicus and Galileo.[4] We will now go over what Kant suggests to us.

The first pathway, he explains, begins with the place that I occupy with my body and brain in the sensible external world and enlarges the connection in which I find myself in "an unending greatness with world after world and system after system: and even more in the unlimited times of their periodic movement, their beginning and their duration". This experience is the pathway of reflection on nature which leads the spirit – admired and lost but nonetheless proud of its awareness of the infinite – beyond the horizon of the visible world.

The second pathway, namely the experience of the moral law, begins with:

> "my invisible, that is to say from consciousness of my personhood, and represents me in a world which has" [*and here the point should be stressed*] "true infinitude, but which only the intellect can penetrate and by which (and thus also at the same time as all these visible worlds) I know myself in a connection, not as is the case there [*in the experience of the starry sky above me*], which is simply accidental, but, rather, which is universal and necessary".

---

[2] Kant, *Werke* VIII, 343 f.. In the preface of the *Anthropologie in Pragmatischer Hinsicht* of the same year (1800) he declared: "the most important object ["der wichtigste Gegenstand"] of all the research in the field of culture is man", cf. Kant, *Werke* VIII, 3.

[3] Kant, *Kritik der praktischen Vernunft*, 289; *The Critique of Practical Reason*. English translation (with my adjustments): cf. http://www.gutenberg.org/dirs/etext04/ikcpr10.txt.

[4] Kant, *Kritik der reinen Vernunft*; *The Critique of Pure Reason*, preface to the second ed. (1787), English translation cf. http://etext.library.adelaide.edu.au/k/kant/immanuel/k16p/k16p2.html.

In the human being the point of intersection between the two experiences or spectacles – however you wish to call them – is striking and revealing of the true infinity, although by participation, of the human spirit, that is to say, of the intellective soul at its centre.

The first spectacle, Kant explains, is that of an innumerable quantity of worlds, whose magnitude, as we are told by contemporary astrophysics, amounts to millions and hundreds of millions of light years (a light year is equal to 9,460 billion kilometres), and is analogous in contrast with that proposed by the microphysics of sub-nuclear particles which enables us to see the duration of a measurement of a hundredth of a thousandth of a billion of a billion of a second and energies of a level of billions of billions of an electron volt. Thus Kant can well say that this spectacle of infinite magnitudes.

> completely annuls my importance as an animal creature, who must return once again to the planet (a mere point in the universe) that matter of which he was formed, after being provided for a short time (without knowing how) of a vital force.

We can say that the extreme nearness that seems to be created between the primary forces of the cosmos and the ultimate particles of matter indicates that man finds himself, as a body, a participant in the creation, of which he, too, in his earthly adventure, is an element and a moment – both in the complex structures of the laboratories of science and in the humble events of daily life.

The second spectacle, instead,

> infinitely raises my spirit in the sense of an intelligence through my personhood whose moral law manifests to me a life that is independent of our being animals and also of the entire living universe, at least as regards what one can understand from what is determined in conformity with the ends of my existence through this moral law, whose indications are not confined to the conditions and the limits of this life, but, rather, which extend to the infinite.[5]

Such for Kant is the grave and irreplaceable task of philosophy as 'practical wisdom' (*Weisheit*): to interpret, on the outside, the truth of being of the presence of greatness in constant expansion of the physical world and life, and, on the inside, the depths of freedom as the truth of being of man. And thus the soul emerges as a subject that is possessed and is presented as the I, as itself, and thus as spirit, beginning with which the self can be said to be an agent and suffering.

The first spectacle or experience is the pathway of science, in particular beginning with the 'Copernican revolution' as Kant himself called it, aware as he was of the new direction taken by Copernicus and Galileo. The other spectacle or experience is the pathway of 'knowing yourself' which Heraclitus glimpsed in the *Logos* and Aristotle identified in the human intellect which was "capable of becoming and doing everything",[6] as an overcoming of the naturalistic physicism of the Pre-Socratics. The two spectacles or experiences intertwine in man.

It follows from these two spectacles or experiences that man's knowledge is not a matter of a single plane or level – that of external observation, explanation, and

---

[5] Ibid.

[6] Aristotle, *De Anima*, III, 5, 430 a 14–16.

experimentation (as a reproduction of phenomena) which is the pathway of modern science. This knowledge develops in the interface between the natural observation of science and the reflective understanding of philosophy. The human being is simultaneously an observable being, like all the beings of nature in which he participates, and a being who interprets himself, who knows himself as Heraclitus and later Socrates had already suggested (a 'self-interpreting being' to employ the definition of Charles Taylor or Paul Ricœur).

This statement on these two objective levels of knowledge that combine in man, the one of the external world which is the object of science and the one inside him, which has the I as its leader, can provide an answer of reconciliation and pacification to the question raised by the status of the human being in the field of knowledge in the age of the predominance of science, as long as, that is, positivist ideology does not claim the right to abolish the border between the sciences of nature and the sciences of man and to annex the latter to the former.

With this spirit we can reconcile a conflict – that connected with the science of genetic mutations or heredity, which, although discovered (and let us not forget the point) by the Augustinian monk Gregor Mendel (1822–1884), was, after Charles Darwin (1809–1882), frequently linked to the theories of evolution. No external limit can be imposed on the hypothesis according to which random variations and given changes have been established and reinforced within the 'narrow corridor of evolution' in order to ensure the survival of a species, and thus of the human species as well. Hitherto we have had historical and perhaps biological evidence, therefore we have something more than a hypothesis, to employ the phrase of Pope John Paul II, in relation to which the experimental sciences must apply greater empirical rigour.

Philosophy, in turn, and not only philosophy but also the social sciences open to the knowledge of biology, must not engage in a battle, which is lost from the outset, to establish the natural facts. Philosophy should ask itself how it can find a meeting point with the scientific point of view, starting from the position according to which the human being is already a speaking, questioning being: a human being, therefore, who has given himself some answers that speak of his domain of freedom in relation to given nature. While the scientist follows the descending order of species and brings out the uncertain, contingent and improbable aspects of the results of evolution in man, the philosopher starts from the self-interpretation of man's intellectual, moral and spiritual situation and ascends back through the course of evolution to the sources of life and of being that man himself is. The starting point can still be the original question, which has always been latent with a sort of self-referentiality of principle: moral law for Kant is what makes the difference; freedom is what Hegel calls "the essence of the spirit".[7]

The human being, discovered and recognised to himself as moral and free, can legitimately ask himself how he arose from animal nature. Thus the approach is retrospective and retraces the chain of mutations and variations. This retrospective approach meets the other, progressive, approach, which descends the river of the

[7] Hegel, *Enzyklopädie*, § 482.

progeny of the human being – man and woman. The two approaches intersect at a point: the birth of a symbolic and spiritual world where moral law and achieved freedom define the humanity of man.

The confusion that has to be avoided lies in the two meanings that can be attributed to the term 'origin': the meaning of genetic or horizontal derivation and the meaning of ontological or vertical foundation. One refers to the origin of species in the succession of space and time beginning with an already originated datum; the other poses the question about the appearance of participated being beginning with the Being by essence. This is the first origin of the being that is the 'passage' of the being from nothing to being which is not properly a passage but rather the primary origin of the being that emerges from nothing thanks to the act of participated being:

> "*Ex hoc quod aliquid est ens per participationem, sequitur quod sit causatum ab alio*", i.e. "from the fact that a thing has being by participation, it follows that it is caused".[8]

Hence the complete formula of the creation as participation (passive in the creature and active in God):

> "*Necesse est dicere omne ens, quod quocumque modo est, a Deo esse*", i.e. "It must be said that every being that in any way is, is from God".[9]

Essential in this vertical 'origin' is the analogical decentring towards the profound, or rather, towards the self of each person, and the analogical recentring upwards, namely towards God. This was also observed by St Thomas in his late work:

> "*Deus est et tu: sed tuum esse est participatum, suum vero essentiale*" i.e. "God is and you [are]: but your being is participated, His is the essential being".[10]

The passage from simple being as an animal creature, to use the phrase of Kant, towards the metaphysical dignity of spiritual being analogous to that of God, is founded on man's dignity as "*forma per se subsistens*", that is, intellective soul, transcendent I, thanks to the direct belonging of the intellective soul to being (*esse*) or to the participated act of being (*actus essendi*).[11] St Thomas is very determined on this point which is the most original point of his anthropology but little known to modern philosophy:

> "when the foundation of matter is removed – as in the spiritual substance and in the human soul –, if any form of a determinate nature remains which subsists of itself but not in matter, it will still be related to its own being (*esse*) as potency is to act. But I do not say, as that

---

[8] St Thomas Aquinas, *Summa Theologica*, I, q. 44, a. 1 ad 1.

[9] Ibid., I, q. 44, a. 1.

[10] Ibid.; *In Psalmum XXXIV*.

[11] "Ipse Deus, qui est esse tantum, est quodammodo species omnium formarum subsistentium quae esse participant et non sunt suum esse", i.e. "God Himself, who is only being, is in a certain way the species of all subsistent forms that participate in being and are not being itself": St Thomas Aquinas, *De potentia*, q. 6, a. 6 ad 5.

> potency which is separable from its act, but as a potency which is always accompanied by its act." He thus concludes that in spiritual creatures as well there is a "composition of potency and act".[12]

The conclusion that can be drawn from this very lofty speculative reflection by St Thomas is that the dignity of being spirit is characterised by Kant, Hegel and others (such as Wittgenstein) after the Galilean revolution in convergence with St Thomas: in modern thought through the transcendentality of knowledge, of freedom, of moral law and of language that have the self as their leader; in St Thomas these transcendentalities, like the self as well, are founded in the act of being and its necessary belonging to the (finite) spirit is obtained by means of the direct participation of God. Therefore, each single subsistence, as Kierkegaard also showed, has his vertical origin as a created person. Thus man is "capable of God", as is rightly observed at the beginning of the Compendium of the Catechism promulgated by Pope Benedict XVI.

[12] "Unde in rebus compositis est considerare duplicem actum, et duplicem potentiam. Nam primo quidem materia est ut potentia respectu formae, et forma est actus eius; et iterum natura constituta ex materia et forma, est ut potentia respectu ipsius esse, in quantum est susceptiva eius. Remoto igitur fundamento materiae, si remaneat aliqua forma determinatae naturae per se subsistens, non in materia, adhuc comparabitur ad suum esse ut potentia ad actum: non dico autem ut potentiam separabilem ab actu, sed quam semper suus actus comitetur. Et hoc modo natura spiritualis substantiae, quae non est composita ex materia et forma, est ut potentia respectu sui esse; et sic in substantia spirituali est compositio potentiae et actus" (*De spiritualibus creaturis*, a. 1 co.). Cf. *Disputed questions on spiritual creatures*, English translation (with my adjustments)cf. http://www.diafrica.org/kenny/CDtexts/QDdeSpirCreat.htm.

Session II: Mathematics – Physics and Cosmology

16 May 2008 – 3:30 p.m.

Chair:
Per Olof Lindblad

Speakers:
Yuri Manin *Truth as a Value and Duty: Lessons of Mathematics*
Bengt Gustafsson *Thoughts on Truth in Physics, Astrophysics and Cosmology*

Panel:
Kai Hauser; Paolo de Bernardis

Per Olof Lindblad:

I would like to ask Yuri Manin, who is professor and director of the Max Planck Institut für Mathematik in Bonn as well as professor at Northwestern University, Evanston Illinois, to deliver his lecture.

Yuri Manin:

I'm very pleased to speak before such a distinguished audience. The title of my talk is 'Truth as a Value and Duty', so I have no doubts about the existence of truth.

Truth as a Value and Duty: Lessons of Mathematics

Yuri Manin:

I Introduction

Imagine that you open your morning newspaper and read the following report:

> Brownsville, AR. A local object partially immersed in a liquid was buoyed upward Tuesday by a force equal to the weight of the liquid displaced by that object, witnesses at the scene reported. As of press time, the object is still maintaining positive buoyancy.

Balzan (ed.), *Truth in Science, the Humanities and Religion: Balzan Symposium 2008*,
DOI 10.1007/978-1-4020-9896-3_3,

In fact, I did read this report in the *ONION*; I have only abridged it to add a Fénéonian touch.

If this meeting had been dedicated to the nature of the comical, one could produce an interesting analysis of the clever silliness of this parody. But as we are preoccupied with truth, I will use it in order to illustrate the differences between the attitudes to truth among practitioners of social sciences and law as exemplified by Bovens and Hartmann (2003) on the one hand, and that of, say, physicists, on the other.

To put it crudely, in social sciences information comes from witnesses; but in what sense was Archimedes' role in his discovery that of a witness, and are the experimental observations generating/supporting a physical theory on an equal footing with the observations of witnesses to a crime scene, or respondents to a poll?

Now, imagine another report, that could have been posted on the website of the Department of Physics of Cambridge University:

> The Cavendish Laboratory News & Features bulletin announced yesterday that a Cavendish student has won a Science, Engineering and Technology award. He managed to measure the constant π with unprecedented precision: $\pi = 3{,}1415925$ ... with an error ± 2 at the last digit.

I must confess right away that I did not read but simply fabricated this spoof in order to stress the further differences between the attitudes towards truth, now held by physicists and by mathematicians respectively.

On the one hand, formally such an announcement would make perfect sense: the *mathematical* constant π can be measured with some precision, in the same way that any *physical* constant such as the speed of light c, or the mass of the electron can be measured. The maximum achievable precision, at least of a 'naive' direct measurement of π, is determined by the degree to which we can approximate ideal Euclidean rigid bodies by real physical ones. The limits to this approximation are set by the atomic structure of matter, and in the final analysis, by quantum effects.

On the other hand, in order to get in principle as many digits of π as one wishes, measurements are not required at all. Instead, one can use one of the many existing formulas/algorithms/software codes and do it on a sheet of paper, a pocket calculator, or a supercomputer. This time the limits of precision are determined by the physical limitations of our calculator: the size of the sheet of paper, memory of computer, construction of the output device, available time ...

What I want to stress now is that π imagined as an infinite sequence of its digits, is not amenable to a 'finite' calculation: even the number of digits of π equal to the number of atoms in the observable Universe, would not exhaust π. Nevertheless, mathematicians speak about π and work with π as if it were a completely well defined entity, graspable in its entirety not only by one exceptional supermind, but by the minds of all trained researchers, never doubting that when they speak of π, they speak about one and the same ideal object, as rigid as if it really exists in some Platonic world.

In fact, one facet of this rigidity can be expressed by a few theorems implying that whatever exact formula, algorithm, or software code we might use to calculate π and whatever precision we choose, we will always get the same result. If we do not, either our formula was wrong, or the calculator made a mistake/there was a bug in the code/output device could not cope with the quantity of information ...

Contemplating this example, we may grasp the meaning of the succinct description of mathematics by Davis and Hersh: "the study of mental objects with reproducible properties".[1]

However, I want to use this example in order to stress that most of the deep mathematical truths are about infinity and infinitary mental constructs rather than experimentally verifiable finitary – and finite – operations, that can be modelled using actual objects of the physical world.

II

... mais je ne le crois pas!!

G. Cantor to R. Dedekind, June 29, 1877

Before Georg Cantor, infinity appeared in mathematical theorems mostly implicitly, through the quantifier 'all' (which also could be only implicit as in most Euclid's theorems).

Cantor proved the first theorem ever in which infinities themselves were objects of consideration. Slightly modernizing his arguments, we can say that he invented two or three mental constructions allowing us to compare sizes (technically, cardinalities) of infinite (in fact, finite as well) sets:

(a) Two sets X, Y have equal cardinalities, symbolically $|X| = |Y|$, if their elements $x \in X$, $y \in Y$ can be joined in pairs (x, y) in such a way that each x is paired with exactly one y and each y with exactly one x.
(b) The cardinality of X is called "less or equal" to that of Y, symbolically $|X| \leq |Y|$, if there is a subset $X' \subset Y$ such that $|X| = |X'|$.

After these two *definitions*, the famous Cantor's *theorem* can be proved in several lines:

(c) The set of all subsets of X, symbolically $P(X)$, has cardinality strictly larger than that of X.

Since we may iterate this construction, forming consecutively $P(P(X))$, $P(P(P(X)))$, ..., we see that *there exists an infinite scale of infinities of growing sizes.*

The proof of (c) consists of two remarks. The first one says that $|X| \leq |P(X)|$, because X can be in a tautological way paired with a part of $P(X)$ consisting of one-element subsets of X.

The second remark is (a remake of) the famous Cantor's diagonal argument, using *reductio ad absurdum*. Imagine that $|X| = |P(X)|$. Then we can pair each $x \in X$ with some $S_x \subset X$ in such a way that any subset $S \subset X$ has the form S_y for some $y \in S$. Choose such a pairing (technically, one-to-one correspondence). Define S = *the set of all x such that $x \notin S_x$.*

This S must be of the form S_y for some $y \in X$, but then both logical possibilities, $y \notin S_y$ and $y \in S_y$ lead to a contradiction, so that the postulated one-to-one correspondence cannot exist.

[1]Davis – Hersh, Mathematical Experience.

Of course, the last key argument goes back to the ancient 'liar's paradox'. It was revived again in a different context in the twentieth century by Tarski and Gödel. Tarski's theorem features the ominous – at least for the purposes of this conference – term "inexpressibility of truth".

In the final analysis, self-referentiality was used to produce several deep mathematical arguments, and this became possible only when the mathematical universe became so extended that the language of mathematics could be embedded into this universe as a part of it. In particular, Leibniz's dream of merging language with meta-language became a reality.

III

> The best test of truth is the power of the thought to get itself accepted in the competition of the market.
>
> Justice Oliver Wendell Holmes, Jr (1919)

When Cantor first presented his diagonal argument in a letter to Dedekind in 1873, it was worded differently and used only to prove that the cardinality of the natural numbers is strictly less than that of the real numbers. The discovery of the proof itself was in a sense hardly more important than the discovery of the definition of what it means, for one infinity to be larger than another one.

As soon as this was achieved, Cantor started thinking about the cardinality of the reals compared with that of the pairs of reals, or, geometrically, sets of points of a curve and of a surface respectively. They turned out to be equal! If we have a pair of numbers (α, β) in $(0, 1)$, Cantor suggested to produce from them the third number $\gamma \in (0, 1)$ by putting decimal digits of α to the odd places, and those of β to the even places. One sees, that vice versa (α, β), can be reconstructed from γ. Dedekind, who was informed by Cantor's letter about this discovery as well, remarked that this does not quite work because some rational numbers have two decimal representations, such as 0.499999 = 0.5000000 ... . Cantor had to spend some time to amend the proof, but this was a minor embarrassment, in comparison with the fascinating novelty of the fact itself:

> Ce que je vous ai communiqué tout récemment est pour moi si inattendu, si nouveau, que je ne pourrai pour ainsi dire pas arriver à une certaine tranquillité d'esprit avant que je n'aie reçu, très honoré ami, votre jugement sur son exactitude. Tant que vous ne m'aurez pas approuvé, je ne puis que dire: je le vois, mais je ne le crois pas!!

as Cantor famously wrote to Dedekind.

This returns us to the basic question on the nature of truth.

We are reminded that the notion of 'truth' is a reification of a certain relationship between humans and *texts/utterances/statements*, the relationship that is called 'belief', 'conviction' or 'faith', and which itself should be analyzed, together with other primary notions invoked in this definition.

In his keynote lecture, Professor Blackburn extensively discussed other relationships of humans to texts, such as scepticism, conservatism, relativism, deflationism.

However, in the long range all of them are secondary in the practice of a researcher in mathematics.

So I will return to truth.

I will skip analysis of the notion of 'humans' and will only sketch what must be said about texts, sources of conviction, and methods of conviction peculiar to mathematics.

Texts: Alfred North Whitehead allegedly said that all of Western philosophy was but a footnote to Plato. The underlying metaphor of such a statement is: "Philosophy is a text", the sum total of all philosophic utterances.

Mathematics decidedly is *not* a text, at least not in the same sense as philosophy. There are no authoritative books or articles to which subsequent generations turn again and again for wisdom. Except for historians, nobody reads Euclid, Newton, Leibniz or Hilbert in order to study geometry, calculus or mathematical logic. The life span of any mathematical paper or book can be years, in the best (and exceptional) case decades. Mathematical wisdom, if not forgotten, lives as an invariant of all its (re)presentations in a permanently self-renewing discourse.

Sources and methods of conviction: mathematical truth is not revealed, and its acceptance is not imposed by any authority.

Moreover, mathematical truth decidedly is not something that can be ascertained, as Justice Oliver Wendell Holmes put it, by "the majority vote of the nation that could lick all the others"[2] or by acceptance "in the competition of the market". In short, it is not a democratic value.

Ideally, the truth of a mathematical statement is ensured by *a proof*, and the ideal picture of a proof is a sequence of elementary arguments whose rules of formation are explicitly laid down before the proof even begins, and ideally are common for all proofs that have been devised and can be devised in future. The admissible starting points of proofs, 'axioms', and terms in which they are formulated, should also be discussed and made explicit.

This ideal picture is so rigid that it can itself become the subject of mathematical study, which was actually performed and led to several remarkable discoveries, technically all related to the effects of merging language with metalanguage and self-referentiality.

Of course, the real life proofs are rendered in a peculiar mixture of a natural language, formulas, motivations, examples. They are much more condensed than imaginary formal proofs. The ways of condensing them are not systematic in any way. We are prone to mistakes, to taking on trust others' results that can be mistaken as well, and to relying upon authority and revelations from our teachers. (All of this should have been discussed together with the notion of 'humans' which I have wisely avoided.)

Moreover, the discovery of truth may, and usually does, involve experimentation, nowadays vast and computer-assisted, false steps, sudden insights and all that which makes mathematical creativity so fascinating for its adepts.

One metaphor of proof is a route, which might be a desert track boring and unimpressive until one finally reaches the oasis of one's destination, or a foot path in green

[2] Peters, *Courting the Abyss*.

hills, exciting and energizing, opening great vistas of unexplored lands and seductive offshoots, leading far away even after the initial destination point has been reached.

IV

> "mismanagement and grief": here you have that enormous distance between cause and effect covered in one line. Just as math preaches how to do it.
>
> J. Brodsky, On "September 1, 1939" by W. H. Auden

Mathematics is most visible to the general public when it posits itself as an applied science, and in this role the notion of mathematical truth acquires distinctly new features. For example, our initial discussion of π as an essentially non-finitary ('irrational') real number becomes pointless; whenever π enters any practical calculation, the first few digits are all that matters.

In a wider context than just applied science, mathematics can be fruitfully conceived as a toolkit containing powerful cognitive devices. I have argued elsewhere[3] that these devices can be roughly divided into three overlapping domains: models, theories, and metaphors. I quote:

> A mathematical *model* describes a certain range of phenomena qualitatively or quantitatively but feels uneasy pretending to be something more.

From Ptolemy's epicycles (describing planetary motions, ca 150) to the Standard Model (describing interactions of elementary particles, ca 1960), quantitative models cling to the observable reality by adjusting numerical values of sometimes dozens of free parameters (≥20 for the Standard Model). Such models can be remarkably precise.

Qualitative models offer insights into *stability/instability*, *attractors* which are limiting states tending to occur independently of initial conditions, *critical phenomena* in complex systems which happen when the system crosses a boundary between two phase states, or two basins of different attractors. [...]

What distinguishes a (mathematically formulated physical) *theory* from a model is primarily its higher aspirations. A modern physical theory generally purports that it would describe the world with absolute precision if only it (the world) consisted of some restricted variety of stuff: massive point particles obeying only the law of gravity; electromagnetic field in a vacuum; and the like. [...]

A recurrent driving force generating theories is a concept of a reality beyond and above the material world, reality which may be grasped only by mathematical tools. From Plato's solids to Galileo's 'language of nature' to quantum superstrings, this psychological attitude can be traced sometimes even if it conflicts with the explicit philosophical positions of the researchers.

A (mathematical) *metaphor*, when it aspires to be a cognitive tool, postulates that some complex range of phenomena might be compared to a mathematical

[3] Manin, *Mathematics as Metaphor*; Manin, 'Mathematical Knowledge'.

construction. The most recent mathematical metaphor I have in mind is Artificial Intelligence (AI). On the one hand, AI is a body of knowledge related to computers and a new, technologically created reality, consisting of hardware, software, Internet etc.. On the other hand, it is a potential model of functioning of biological brains and minds. In its entirety, it has not reached the status of a model: we have no systematic, coherent and extensive list of correspondences between chips and neurons, computer algorithms and brain algorithms. But we can and do use our extensive knowledge of algorithms and computers (because they were created by us) to generate educated guesses about structure and function of the central neural system. [...].

A mathematical theory is an invitation to build applicable models. A mathematical metaphor is an invitation to ponder upon what we know.[4]

As an aside, let us note that George Lakoff's *definition* of poetic metaphors such as "love is a journey"[5] is itself expressed as a mathematical metaphor using the characteristic Cantor-Bourbaki mental images and vocabulary:

> More technically, the metaphor can be understood as a mapping (in the mathematical sense) from a source domain (in this case, journeys) to a target domain (in this case, love). The mapping is tightly structured. There are ontological correspondences, according to which entities in the domain of love (e. g. the lovers, their common goals, their difficulties, the love relationship, etc.) correspond systematically to entities in the domain of a journey (the travellers, the vehicle, destinations, etc.).

When a mathematical construction is used as a cognitive tool, the discussion of truth becomes loaded with new meanings: a model, a theory or a metaphor must be true to a certain reality, more tangible and real than the Platonic 'reality' of pure mathematics. In fact, philosophers of science routinely discussed truth precisely in this context. Karl Popper's vision of scientific theories in terms of falsifiability (versus verifiability) is quite appropriate in the context of highly mathematicised theories as well.

What I want to stress here, however, is one aspect of contemporary mathematical models which is historically very recent. Namely, models are more and more widely used as 'black boxes' with hidden computerized input procedures, and oracular outputs prescribing behaviour of human users.

Mary Poovey,[6] discussing from this viewpoint financial markets, remarks in her insightful essay that what she calls "representations", basically computerized book-keeping or the numbers a trader enters in a computer, tend to replace the actual exchange of cash or commodities:

> This conflation of representation and exchange has all kinds of material effects, [...] for when representation can influence or take the place of exchanges, the values at stake become notional too: they can grow exponentially or collapse at the stroke of key.

In fact, actions of traders, banks, hedge funds and alike are to a considerable degree determined by the statistical models of financial markets encoded in the software of

[4]Manin, 'Mathematical Knowledge', 14–15.

[5]Lakoff, *Contemporary Theory*.

[6]Poovey, 'Can Numbers Ensure Honesty?'

their computers. These models thus become a hidden and highly influential part of the actions, our computerized 'collective unconscious'. As such, they cannot even be judged according to the usual criteria of choosing models which better reflect the behaviour of a process being modelled. They are part of any such process.

What becomes more essential than their empirical adequacy, is, for example, their stabilizing or destabilizing potential. Risk management assuming mild variability and small risks can collapse when a disaster occurs, ruining many participants of the game; risk management based upon models that use pessimistic 'Lévy distributions' rather than omnipresent Gaussians paradoxically tends to flatten the shock waves and thus to avoid major disasters.[7]

V

> There have been dramatic changes in the way in which the motion of the crowd is modelled in recent years.
>
> R.R. Clements and R.L. Hughes, 'Mathematical Modelling of a Mediaeval Battle'

When in the twentieth century mathematicians got involved in heated discussions about the so called "Crisis in Foundations of Mathematics", several issues were intermingled.

Philosophically-minded logicians and professional philosophers were engaged with the nature and accessibility of mathematical truth (and reliability of our mental tools used in the process of acquiring it).

Logicists (finitists, formalists, intuitionists) were elaborating severe normative prescriptions trying to outlaw dangerous mental experiments with infinity, non-constructivity and *reductio ad absurdum.*

For a working mathematician, when he/she is concerned at all, 'foundations' is simply a general term for the historically variable set of rules and principles of organization of the body of mathematical knowledge, both existing and being created. From this viewpoint, the most influential foundational achievement in the twentieth century was an ambitious project of the Bourbaki group, building all mathematics, including logic, around set-theoretical 'structures' and making Cantor's language of sets a common vernacular of algebraists, geometers, probabilists and all other practitioners of our trade. These days, this vernacular, with all its vocabulary and ingrained mental habits, is being slowly replaced by the languages of category theory and homotopy theory and their higher extensions. Respectively, the basic 'left-brain' intuition of sets, composed of distinguishable elements, is giving way to a new, more 'right brain' basic intuition dealing with space-like and continuous primary images, both deformable and deforming.

In the Western ethnomathematics, truth is best understood as a central value, ever to be pursued, rather than anything achieved. Practical efficiency, authority, success in competition, faith, all other clashing values must recede in the mind of a mathematician when he or she sets down to do their job.

---

[7] Mandelbrot – Hudson, *(Mis)behavior of Markets.*

The most interesting intracultural interactions of mathematics such as symbolized by this conference are as well those that are not direct but rather proceed with the mediation of value systems.

Coda

Every 4 years, mathematicians from all over the world meet at the International Congresses (ICM), to discuss whatever interesting developments happened recently in their domains of expertise. One of the traditions of these Congresses is a series of lectures for general public.

In 1998, our Congress met in Berlin, and Hans Magnus Enzensberger, the renowned poet and essayist, deeply interested in mathematics, spoke about 'Zugbrücke außer Betrieb: die Mathematik im Jenseits der Kultur': the drawbridge to the castle of mathematics is out of service. The main concern of his talk was a deplorable lack of mathematical culture and communication between the general public and mathematicians, leading to alienation and mutual mistrust.

At the end of his talk Enzensberger[8] quotes an imaginary dialogue from Stewart (1987), where a mathematician is chatting with a fictional layman 'Seamus Android':

Mathematician: "It's one of the most important discoveries of the last decade!"

Android: "Can you *explain* it in words ordinary mortals can understand?"

Mathematician: "Look, buster, if ordinary mortals could understand it, you wouldn't need mathematicians to do the job for you, right? You can't get a feeling for what's going on without understanding the technical details. How can I talk about manifolds without mentioning that the theorems only work if the manifolds are finite-dimensional para-compact Hausdorff with empty boundary?"

Android: "Lie a bit."

Mathematician: "Oh, but I couldn't do that!"

Android: "Why not? Everybody *else* does."

And here I must play God and say to both Android and Mathematician: "Oh, no! Don't lie – because everybody else does."

Panel and General Discussion

Kai Hauser:

I would like to ask Professor Manin three questions about his very stimulating presentation: The first question concerns your remark in the beginning, that truth is about infinity and that mathematics is about infinitary mental constructs. Somewhat later you mentioned that Cantor invented mental constructions, for example to compare the sizes of infinite sets and to make distinctions in the infinite.

[8] Enzensberger, *Drawbridge Up*.

My question is this: what exactly should we envisage under mental construction here? Is it a process that occurs in my mind, or your mind, or maybe in all the minds of all mathematicians uniformly, an expression of some sort of biological or neurological uniformity? Or perhaps, is what you mean a process that is carried out in the mind of some idealized subject only? After all, the intended meaning of mathematical statements is that they are mind-independent – they transcend all minds that can grasp them. How do you see this?

The second question is food for thought rather than a real question: It concerns your characterization of truth as a reification of a certain relationship between humans and texts. But I would rather argue as follows: if you think along the lines of the traditional formula, *adaequatio rei et intellectus*, then truth concerns the relationship between propositions and reality. Maybe what you were trying to say is that now the focus has shifted to another notion which hasn't been explicitly discussed in this meeting yet, but which I think is of central importance in all investigations into truth, namely the notion of evidence. Because here, the subjective plays a role. In my view, evidence is something that lies between the subjective and the objective. And I note that our ideas about what counts as evidence of truth seem to be much more concrete than our ideas about the nature of truth itself.

I come to my final question. Towards the end of your talk, you said, a mathematical statement is ensured by a proof. And then you mentioned another concept: the starting points of proofs are axioms. What about the truth of the axioms? Do you think the axioms should be taken at face value? Are they literally true? And also, do you think there is any ultimate significance, from the epistemological point of view, in distinguishing between theorems and axioms?

Yuri Manin:

First of all, my task here at this meeting, as I understood it, was to express feelings and notions with which working mathematicians could probably agree, and feelings and notions relating to truth in mathematics. I didn't want to touch upon philosophy, or, whatever else is relevant to this general notion. So, when I said that truth is a reification of a certain relationship between humans and texts, I wanted to say that if there are no texts, you cannot even start talking about truth. You need some text. It may be a book, it may be a body of knowledge written in any other form. But you need a text, first of all.

Kai Hauser:

What about $2 + 2 = 4$?

Yuri Manin:

I was once asked to produce a public talk, in mathematics, but understandable for everybody. Of course my first thought was, ok, I'll try to explain why $2 + 2$ is always

4. But then I got stuck by the word 'always'. Always means that I should include time. When I include time, I should express that, actually, I speak about a certain conservation law. Conservation law, as we understand it now, is essentially a property of a certain thing which is functioning in the space-time domain: there is an action, a classical motion. So I decided that this was already too difficult, and so my actual talk was devoted to the question 'What is 2?' I could have done this here too, but I would take another hour.

Of course, I could not avoid touching upon philosophical issues at certain points, but as a mathematician and, a little bit, a scientist, I am pretty arrogant with respect to philosophy, and my motto is, when philosophy is poetry, the question of its truth is irrelevant; when philosophy is not poetry, it is neither true nor false, it is superficial.

Question:

I wanted to ask Professor Manin to what extent are mathematical truths out there waiting to be discovered, or to what extent are they constructs of human endeavour?

Yuri Manin:

Basically, this is an old Platonic problem, whether we discover mathematics or invent it. And of course one cannot answer it. But with what I said about philosophy, I would say that, emotionally, I am very much for Plato's picture, and I of course imagine the existence of this gold that we discover and not invent. Therefore, mathematical truths are waiting for us to be discovered. If we are too late, they will live without us. But this is not a rational argument, it's just what I feel.

Question (Simon Blackburn):

I am a bit worried about that last remark, if I may say so. If you take his great book *Foundations of Arithmetics*, the mathematician Frege started with, basically, a precursor or an elaboration of Cantor's account of number. But then he goes on to do quite a lot of philosophy. Do you think his book just went superficial at that point? It's not very poetic.

Yuri Manin:

Yes, in fact I do. There was also a lot of philosophy motivated by mathematics produced in the first third of the twentieth century, and the deep part of it became a part of mathematics, and the superficial part is just forgotten.

Question:

You said there is a correspondence between mathematics and the natural. What is the reason for this correspondence?

Yuri Manin:

It is as indirect a correspondence as that between human language and nature. One could produce statements that, first, are meaningful in terms of their practical or technological or even philosophical interpretation, then, one can ask about statements whether they are mathematically true or not. And if they are mathematically true, and if the consequences of this can be experimentally verified, then mathematics describe this part of nature. Otherwise, it is a cognitive tool, as I said. It's not something that maps into nature one-to-one.

Question:

It's quite amazing that humans, who are just another animal, discovered these rich mathematics whereas dogs or horses haven't. Two questions: would any other sufficiently developed intelligence develop the same kind of mathematics, and if they were even more developed, would they be doing something very different which we could never even grasp?

Yuri Manin:

Well, that was the question I asked myself when I was in my teens – is mathematics on Mars the same as here? And of course it's just a speculation, but I will say that yes, intrinsically it would be the same, although for example, they could have started inventing computers at much earlier stages than we did, and then much more mathematics would become interiorized in computers rather than in Martians' brains. This I can imagine. But the sum total, I think it would be basically the same. Not quite the same: isomorphic to human mathematics.

Question:

Why is mathematical truth beauty?

Yuri Manin:

Because we feel it as beauty! Beauty is also a relationship between us and something.

Per Olof Lindblad:

With mathematics on Mars, we are approaching astronomy. So I call on Professor Bengt Gustafsson, who is professor in astrophysics at the University of Uppsala, to discuss some thoughts about truth in physics, astronomy and cosmology.

Bengt Gustafsson:

The organizers were brave enough to ask me to give a talk on this giant subject: just a simple astronomer dealing only with the universe. But I got some encouragement by studying the seal of our university: *Gratiae veritas naturae*. So maybe it's part of my job.

Thoughts on Truth in Physics, Astrophysics and Cosmology[9]

Bengt Gustafsson:

When Riccardo Giacconi came to Stockholm in December 2002 to get his Nobel Prize in Physics "for his pioneering contributions to astrophysics, which have led to the discovery of cosmic X-ray sources", to cite the prize motivation, he was asked to sign a poster at the Royal Swedish Academy of Sciences. Such posters are issued by the Academy every year and spread in a great number of copies to schools and universities all over the world. On this poster, celebrating the physics prizes of 2002, there was a picture of the X-ray source Cygnus X1, which is a binary system believed to contain a black hole. Under the picture one could read that Giacconi's discovery of X-ray sources like Cygnus X1 is supposed to lead to the eventual discovery of black holes. Giacconi refused to sign. After some persuasion he grabbed the pen, crossed out some words in the poster text, and wrote some others, signed the poster and left. When we scrutinized the modified poster text we read that Giacconi's discovery of X-ray sources had led to the discovery of black holes. For Giacconi, black holes were certain to exist, he had discovered their presence in the real world.

Physicists and astronomers have certainly different attitudes toward philosophical questions, on how knowledge can and should be acquired, and if knowledge is about a real world, existing independently of us, or just constructions to help us orient ourselves and interact with each other and our surroundings. Is there any physical truth independent of the physics community? But the act by which Riccardo Giacconi modified the poster did probably not reflect some basic philosophical difference in

---

[9]Thanks are due to the organizers of this interesting Symposium for inviting me to take part in it, and to the editors of this volume for their patience. I also wish to thank Nils Bergvall, Ulf Danielsson, Gösta Ekspong, Gunnar Ingelman and in particular Otto Sibum, for inspiration, suggestions and helpful discussions. I hope I have been able to give a reasonably fair presentation of thoughts which are due to them.

opinion on the question, at what moment a hypothesis, like that of black holes, has such a strong empirical support that it should be regarded as reality. Almost all contemporary scientists tend to regard scientific knowledge, or at least theories, as provisional. However, the issue is on which grounds we should regard a theory a provisionally correct description of the world.

It is interesting to follow this process in physics and astronomy, when a hypothesis changes character from being mere conjecture to when it reaches the status of provisional truth. A common feature in these developments has been the key significance of certain decisive experiments. I shall here take three glorious examples of this emergence of provisional truth, and ask if they have anything in common.

But first, I have to make some comments about *measurements*. The role the art of exact measurements has played in the development of physics is a fascinating story. There were early pioneers, such as the remarkable Ibn al-Haytham from Basra, present Iraq, who in the tenth and eleventh centuries did accurate optical experiments and concluded that "truth is difficult to find, and the road to it is rough". But the intensive development of exact and controlled measurements took place in the seventeenth century. In a sense, the concept of the scientific 'fact' is an invention of that century. The fact became very much a measure. The laboratory, with air pumps, vacuum chambers, barometers and thermometers was the new place in which scientific factual knowledge was produced. The step from sensual experience to knowledge that could be generalized about a natural order was non-trivial. Could one rely on artificial experiments at all? Can instruments really replace our own senses in producing facts – yes, even better ones than our unreliable senses suggest? Gradually, precision measurement became the key technique to secure reliable knowledge. And this was grounded in a special view of reality among the scientists involved. Steven Shapin argues, in the case of Robert Boyle, that only members of this community sharing this view of reality were able to judge when 'reasonable agreement' was really achieved in laboratory experiments. Shapin continues: "This disciplining of experience thus also contained a social order, and undisciplined experience soon lost its status and became regarded of no use". The idea that ever more accurate measurement was the basis of scientific truth developed further during the eighteenth century. It was, for natural reasons, closely tied to various handicrafts, technology and instrument making. As the historian of science Otto Sibum says:

> Around 1800, precision instruments were employed in various cultural fields of production, but the key motive for their use was the control and regulation of society and not – as one might expect – the mathematicians' ambition to take over subjects of natural philosophy.

So, it was driven by navigation and artillery needs, land surveying, mining, steam engine technology, etc.

Now to the first example, about which we heard earlier today when Professor Severino talked about *the energy principle*. Related ideas – about quantities that were considered to be conserved – appeared early, were based on rather intuitive and sometimes vitalistic grounds, and may be traced through a meandering history from Thales to Leibniz. These ideas were further developed and substantiated by physicists, chemists and engineers in the eighteenth century. One interesting example is the remarkable figure Benjamin Thompson, for a time war minister of Bavaria and known for useful inventions like the Rumford stove and a coffee drip pot. But he was also boring cannons for the military and then studied the developed heat as a function of the action. In the 1830s, the formulations of the principle were getting close to our own versions. The nature of heat was, however, still unclear. That heat could also be regarded as one of the transformable energy forms was suggested, on interesting grounds, by the German physician Julius Robert von Mayer in 1842, but he was an outsider and not generally accepted by the physics community. The firm proof of the conservation principle was instead, to a very important degree, the fruit of the nineteenth century experimental physics culture. James Prescott Joule's work, which originated from his brewery background, was of decisive significance. Otto Sibum (whom I follow closely in this presentation of Joule's work and of the way the conservation principle was established) has noted that brewing culture had its own particular demands of precision measurement for describing phenomena and their relations quantitatively. Joule tried to gain entrance into the gentlemen's science circuits in Britain with his accurate measurements of the transformation of mechanical action to heat using the famous paddle machine. His work focused on measuring a key quantity: the conversion quantity between mechanical work and heat energy, the mechanical heat equivalent. Joule's conclusion from the experiments that "friction consists in the conversion of force into heat" met with hard resistance. The nature of heat was still a philosophically disputed issue.

Joule himself at first did not consider his results as proof of a general principle of energy conservation. Joule cited the astronomer William Herschel who had said that "science has not greater bane than hasty generalisation". Joule's work, as well as the work of others, studying accurately and in detail the energy budget when mechanical, chemical, electrical energy were converted into heat, gradually led him to the conviction that energy was indeed conserved. A part of this development was the construction by Joule and his collaborators of very precise thermometers. The American physicist and engineer Henry A. Rowland got deeply impressed and wrote: "... about one thing we may be certain, that when the scientific millennium is reached [...] only physicists of precision remain". A famous expression for this measuring culture is a statement by Sir William Thomson, Lord Kelvin:

> When you can measure what you are speaking about, and express it in numbers, you know something about it; but when you cannot measure it, when you cannot express it in numbers, your knowledge of it is of a meagre and unsatisfactory kind; it may be the beginning of knowledge, but you have scarcely, in your thoughts, advanced it to the stage of science.

Otto Sibum has presented this development in a fascinating article, which has been my major source for the present discussion. It has the title 'The Number of the Century. A History of a Scientific Fact'. Henry A. Rowland characterized the mechanical heat equivalent as "One of the most important constants of nature". And when the nineteenth century was summarized, his colleague T.C. Mendenhall said:

> The mechanical equivalent of heat, the number of units of work necessary to raise 1 pound of water 1 degree in temperature, has, with very much reason, been called the golden number of that century.

The fundamental significance of the mechanical equivalent of heat, a single number, is not generally held in admiration by today's physicists. In fact, already soon after Mendenhall Max Planck regarded it as just a conversion factor, a trivial consequence of energy conservation which to Planck was a fundamental principle of physics in a reality which existed independently of us and, just because of that, the principle was not verifiable in a fundamental way by any of our measurements. Even if many of us may regard the mechanical equivalent of heat as a fairly trivial conversion factor, the idealistic attitude of Planck is also not commonly shared among physicists. Energy conservation is indeed a powerful principle, e.g., as was demonstrated when the neutrino particle was discovered, which was first invented to 'save' the principle. That does not exclude, however, that energy conservation may be broken, sometimes or somewhere. When somebody makes such claims, however, physicists tend to suspect yet another *perpetuum mobile* enthusiast, and are not willing to give in easily. Although all our theories are provisional, we feel that the energy principle, or the First Law of Thermodynamics, is unusually solid.

A consequence of the work of Joule and others on 'energy physics', pointed out by Sibum, was however that 'energy' changed character from a fairly abstract concept to something which gradually gained status as 'a thing'. The influential German physicist Friedrich Kohlrausch was second director of the Physikalisch-Technische Reichsanstalt in the late nineteenth century and had to advise on practical matters, e.g., in law cases when electricity was tapped during dark evenings by secret wires in order to light flats. Could that be compared with stealing apples? Kohlrausch vividly argued that work and energy indeed were "Sachen" and thus could be stolen. Today, energy is bought and sold on markets, it is transformed and adjusted to the needs and purses of the customers, it is labelled "environmental friendly" or "a commodity of national interest". It seems trivial to say that energy has become real.

Now to our next example, *the quarks*. I spent a few years in the late 1960s at the Niels Bohr Institute in Copenhagen. One day in 1970, one of our professors entered the legendary lunchroom of the Institute. He was excited, his hair in Einsteinian disarray. He had just arrived back from an international high-energy physics conference in Kiev. And he shouted all over the room, "Now they exist! They have seen the quarks, finally!"

Quarks were originally developed from a classification scheme to put some order to the rich zoo of elementary particles, in particular the hadrons such as protons and neutrons. The scheme was originally suggested in 1961 by Murray Gell-Mann and Kazuhiko Nishijima, and the quarks entered the scene in papers by Gell-Mann and independently, George Zweig. The noun 'quark' was given by Gell-Mann, after an enigmatic passage in James Joyce's novel, *Finnegan's Wake*. The basic arguments behind the proposal of the quarks were, as so often in elementary-particle physics, aesthetical – arguments of symmetry. A beautiful order among the hadrons could be established if a global flavour SU(3) symmetry existed. This suggested the hadrons to be composed by smaller particles, with charge units of 2/3 and 1/3 in units of the proton charge.

The use of symmetry and aesthetic arguments was not very mysterious. One could argue that economy was a basic motivation for establishing simple and straight taxonomy for the complex zoo. Also, general experience among the many theoretical physicists made them trust that the results of physical measurements could be organized, described and even predicted in beautiful mathematical theories, and this caused some wonder. A famous expression of this amazement is the essay by Eugene Wigner from 1960 on 'The unreasonable effectiveness of mathematics in the natural sciences'. The reason why mathematical tools are so effective, and may even lead to predictions of qualitatively new physical phenomena like the anti-particles, the black holes or the quarks, is probably still not understood. Also the symmetry invoked for the quarks led to a prediction of a new particle, $\Omega-$, discovered in 1964, which was considered to give strong support to the quark model. Several different competing ideas existed also about the inner world of the hadrons. And they had strong supporters too, in particular after free quarks had been sought for in vain in numerous accelerator investigations and studies of cosmic ray particles.

The quarks could be fitted rather conveniently into the Standard Model of Elementary Particles that Steven Weinberg proposed in 1967. He had, together with Abdus Salam and Sheldon Glashow, succeeded in unifying the electromagnetic and weak nuclear interactions into one theory, and now also the strong nuclear interaction could be included. Weinberg is a clear proponent of the almost Platonian, or Pythagorean, trust that mathematical consistency and beauty prevails in nature. He ended his Nobel Prize lecture from 1979 thus:

> I suppose that I tend to be optimistic about the future of physics. And nothing makes me more optimistic than the discovery of broken symmetries. In the seventh book of the *Republic*, Plato describes prisoners who are chained in a cave and can see only shadows that things outside cast on the cave wall. When released from the cave at first their eyes hurt, and for a while they think that the shadows they saw in the cave are more real than the objects they now see. But eventually their vision clears, and they can understand how beautiful the real world is. We are in such a cave, imprisoned by the limitations on the sorts of experiments we can do. In particular, we can study matter only at relatively low temperatures, where symmetries are likely to be spontaneously broken, so that nature does not appear very simple or unified. We have not been able to get out of this cave, but by looking long and hard at the shadows on the cave wall, we can at least make out the shapes of symmetries, which though broken, are exact principles governing all phenomena, expressions of the beauty of the world outside.

But did these quarks exist in reality, or were they to be seen just as mathematical constructs, of no real reality except for among Platonists? Gell-Mann himself was asked this question, against the background of the failure of the first attempts to find free quarks. He was uncertain: "Such particles [quarks] presumably are not real but we may use them in our field theory anyway."

In California the Stanford Linear Accelerator (SLAC) was just being built. It was 2 miles long, the largest and most expensive accelerator ever built. It was able to accelerate electrons to energies up to 20 GeV, which was several orders of magnitude more than had been used to demonstrate the existence of the atomic nuclei in the early century, and in the 1950s to explore the inner structure of the nuclei. But the planned approach was now analogous to that of those pioneering studies, to shoot very energetic electrons onto the hadrons, the protons and neutrons, to see how the electrons were scattered away and transformed. In 1967 the experiments started, led by Jerome Friedman, Henry Kendall and Richard Taylor and in 1968 the first astonishing results were known. The scattering was much more efficient than thought, and indicated intricate sub-structure inside the nuclear particles. A series of further experiments with different energies and different scattering angles, experiments which were partly interpreted by Richard Feynman and James Daniel Bjorken, proved that there are three points of deflection in protons and neutrons while there are two in mesons, just as it had been suggested by Gell-Mann, who had stated that protons and neutrons consisted of three quarks, and mesons of two. Also the charges that were found later were as suggested by the Standard model. Even other ingredients, the glue-particles – gluons, that hold the particles together in the hadrons – could be traced. This caused the excitement of the professor in the lunch room in Copenhagen.

Some years later the findings from SLAC had been verified by other experiments, and the reason why free quarks were never found was understood – the quarks cannot be pulled out of the nucleons, while they have free mobility inside, as *the theory of asymptotic freedom* has demonstrated. And the standard model has predicted two additional families of quarks which subsequently have been found experimentally, and even the quark pair called charmonium, composed by one charm quark (these particles have very charming names) and one anti-charm quark. The theory behind the Standard Model is not complete, since particle masses and strengths of interactions must be put in by hand, and gravity is not included either. But quarks are generally considered to exist and now also appear in elementary school books.

When their proposer Murray Gell-Mann later commented on his earlier statement that "they probably did not exist" he said:

> That is baloney. I have explained so many times that I believed from the beginning that quarks were confined inside objects like neutrons and protons, and in my early papers on quarks I described how they could be confined [...] which is what we believe today to be correct. Unfortunately, I referred to confined quarks as 'fictitious', meaning that they could not emerge to be utilized for applications...

My next example is the discovery of the *Big Bang*. That moves us from the field of experimental physics to a philosophically very different situation. Unlike physicists, astronomers can only rarely conduct any meaningful experiments. They are left to observe instead, which means that they have to take their composite objects – stars, galaxies – as they are, they cannot prepare them or provoke them to see how they react. And the cosmologists are even worse off. They cannot work like normal astronomers, comparing different stars, different galaxies or planets, just as the biologist compares different plants or animals or an archaeologist compares different cultures. No, the cosmologist has just one universe, one specimen, to investigate, and he tries to reconstruct its history and its origin. It is a challenge to make science out of that, and not long ago cosmology was indeed not regarded as a science by many of my colleagues. The success of the Big Bang cosmology has changed this situation.

One may trace the origin of this grand creation myth of our times back to the Jewish-Christian mythology or even more ancient ideas. The modern history of the Big Bang began in 1917 with the Dutch astronomer Willem de Sitter, who found a solution to Einstein's field equations of General Relativity for the Universe as a whole, which he, to make it simple, assumed to be empty. His cosmological model could expand with exponentially increasing speed. In a sense Einstein's theory had thus predicted that the Universe might expand, in a time when most scientists, including Einstein himself, thought that it was basically steady and had been so forever. Some years later the Russian mathematician Alexander Friedman showed that Einstein's equations could be solved also for cosmological models with matter in them, and again found solutions that contracted or expanded. But these findings were not much noted. In 1927 similar results were found independently by abbé Georges Lemaître, who was theoretical physics professor in Louvain. De Sitter's and Friedman's findings were purely theoretical, and so were those of Lemaître. It seems, however, that the latter was aware of an enigmatic result that several astronomers had found already ten years before: that a clear majority of the galaxies showed spectra with red-shifted spectral lines, suggesting that they had systematic motions away from us. It was, however, the American Edwin Hubble with his collaborator Milton Humason at Mt Wilson Observatory who would establish the tendency clearly: Hubble plotted the redshift as a function of distance and found a linear relation, the Hubble relation. This fundamental discovery was published in 1929, obviously without any knowledge of the previous theoretical work by De Sitter, Friedman and Lemaître. And only gradually the discovery was interpreted in terms of a general expansion of the Universe. Hubble himself, in a famous book from 1936, wrote: "Conclusions from data now available are rather dubious." Abbé Lemaître, however, who now became recognized as a theorist who had predicted Hubble's observations of the expanding universe, drew the velocity arrows of the galaxies backwards in time and soon elaborated his views on how it all began, namely from a compressed state which he called "l'hypothese de l'atom primitif" (the hypothesis of the primeval atom). An initial "creation-like" event must have started it all "on a day without yesterday" the abbé conjectured. He also phrased it as "the cosmic egg exploding at the moment of the creation". Einstein and the leading British astrophysicist Arthur Eddington supported Lemaître's work, but both disliked his idea about the superdense beginning. Lemaître became old enough to experience its verification, however, by the discovery of the Cosmic Microwave Background in the mid 1960s.

Long before then, the Russian nuclear physicist George Gamow had started his work on cosmology. He was a student of Alexander Friedman but left the Soviet Union early for Germany, Denmark and the United States. His hope was to prove that the chemical elements could have been formed by nuclear reactions in the hot and dense early universe. Unlike Lemaître, however, he was not a religious character. Instead he liked to joke or ridicule authorities, ecclesiastical as well as scientific. He and his student Ralph Alpher showed in a paper from 1948 that much of the helium in the universe could have been formed in this way, but it was difficult to explain the heavier elements. Inspired by Gamow Ralph Alpher and Robert Herman also predicted that it should still be possible to trace an afterglow of the thermal radiation in the early universe as a thermal microwave radiation at about 5° above the absolute zero-point. It took time before this prediction attracted attention, perhaps partly because of Gamow's not fully serious manners. He adopted the selling term "Big Bang" for his cosmology from a critical remark by Fred Hoyle, with whom he discussed the theory in a BBC radio programme in 1950.

Hoyle became the foremost proponent for an alternative theory to the Big Bang. In his and his collaborators' 'steady-state cosmology' from 1948, 'the perfect cosmological principle' reigns, meaning that the Universe is roughly the same at any place and at any time. Thus, the mean density of the universe was assumed to be constant in spite of its general expansion. To accomplish this, matter was supposed to be continuously formed, at a gentle rate, in between the stars and galaxies to compensate for the thinning due to the expansion. The very idea of the steady state alternative was that the Universe should not change basically. However, early in the 1960s it was found that the density of radio galaxies and quasars was larger at long distance, that is long ago since we look backwards in time when we look out. This was a blow to the steady-state idea. And then came the discovery, by chance, of the Cosmic Microwave Background by Arno Penzias and Robert Woodrow Wilson in 1965. These authors, who were trying to improve satellite communication systems at Bell Laboratories, did not know at all that the microwave background had been predicted long ago, nor did they first understand that they had stumbled across a major discovery, the relic from the Big Bang. When this was understood, the support for its competitor, the steady-state cosmology, gradually declined.

For 25 more years, there remained some hope among those who had an aesthetically or philosophically based feeling for the steady-state hypothesis: the observations from ground of the spectral distribution of the microwave background radiation did really not look like the predicted black-body radiation, the smooth Planck curve. New evidence for the Big Bang appeared, such as observations of a uniform lithium abundance in the atmospheres of certain types of old stars which were thought to reflect the original composition of the gas from which they had been formed. And these small amounts of lithium agreed reasonably well with those predicted from nucleo-synthesis in the Big Bang fireball. The moment of truth finally came when the Cosmic Background Explorer (COBE) satellite had been up for just a few days in 1990. Its FIRAS instrument measured the spectrum of the microwave background and delivered a perfect Planck curve, in fact better than any that could be created in laboratory experiments. From that moment 18 years ago, Big Bang existed among us.

In the last decades, the observational cosmology has developed impressively, producing much better data for the angular distribution of the cosmic microwave background and the spatial distribution of galaxies. This has led to further confirmation of the Big Bang theory and to the filling in of many details in the overall picture. Some observed enigmas have also become obvious: the nature of the as yet not well known 'dark matter' which provides about five times more gravitational mass density than all normal and visible matter. Still more problematic is the 'dark energy', a mechanism (whatever it is) that keeps accelerating the expansion of the universe since about 5 billion years ago. These two unknown phenomena dominate the overall dynamics of our universe today. The physics of the probable, very rapid (exponential) expansion in the very early epoch in the history of the universe (within the first 10^{-30} s or so!), 'the inflation era' also remains to be understood. The very first moments which can be discussed today, which are close to the Planck time (about 10^{-43} s), require a still missing consistent quantum-mechanical theory of gravity. Whether the fascinating super-string theory will manage to fill in this gap in the cosmological puzzle remains to be seen. Existing attempts in this direction lead to new interesting philosophical problems such as the suggestions that a 'landscape' of universes (maybe amounting to 10^{500} different worlds!) in a multidimensional space, with different 'laws of nature' in different universes, could exist. Our universe would then perhaps be one of the few 'habitable' ones, and its physical properties would then relate to our existence.

I shall now end discussing some common features that these three different developments, of the energy principle, the quark model and the Big Bang cosmology, show. First of all, all three cases have a background in speculation and theory. So, for instance, in the principle of energy conservation, this background is fairly obscure and partially not very physical. The quark model was suggested by contemporary leading physicists for theoretical and aesthetical reasons to bring order and symmetry into the wealth of the empirically found so-called elementary particles. The Big Bang was predicted on the basis of a most beautiful theory, Einstein's theory of General Relativity, and first against Einstein's instincts. The leading figure in the early history of the Big Bang hypothesis was Lemaître, a priest and physicist who clearly saw the correspondence between biblical creation and his own idea.

In all these cases, the experiments played a key role. In the energy conservation case, the meticulous work by Joule and collaborators, which has been described as "a pragmatic empirical deductive method", is of decisive importance. This work was driven by a culture where the significance of experimentation and handicraft skills in experimentation were held in very high esteem, leading to data that in themselves served as symbols of real truth to the community. Truth was a number. The very resource demanding key experiments proving the quark model could, due to happy circumstances, be carried out in the decade after its appearance as a theoretical hypothesis. The experimental support for the Big Bang model was accumulated, but in a partly circumstantial way, because some early theoretical predictions from the model were not widely regarded as a basis for tests to be checked. In all these cases however, there were some key experiments that, to a broad scientific community, demonstrated the inevitable truth. From then on, the issue was settled. The energy principle, the quark model, the Big Bang were found true due to such, I would say, almost emblematic demonstration experiments. At least they were found provisionally true, and further than that the ambition of physicists and astronomers does not extend.

However, when going through these different stories, and looking at the scientists involved in them and the scientists commenting on contemporary developments, it is very clear that the philosophical epistemic positions were different, and widely so. There were naïve realists among them, there were idealists, and yes, even Platonists. There were pragmatists, and there were, most probably, even if they did not call themselves that, deflationists. For some, the laws of nature were economical man-made constructions to make it possible to handle nature. For others, they rather were revealing deep truths in our own mental inner self. And that is still the position. Any deeper discussion between different active physicists discloses widely different ideas about what truth in science is, and indeed what science is and how it should be understood. And this then leads to a paradox: a life in scientific committees and research councils has demonstrated to me that there is a remarkable consensus among physicists and astronomers, in spite of these different positions, on what is good science. The consensus is also shown in the possibility of attaining unanimity when it comes to selecting individuals to positions or prizes. Why this consensus, when the basic ideas are so different?

Several philosophical and sociological explanations why a consensus is maintained can be envisaged, and I shall mention some of them here. Maybe a general pragmatism of physical sciences has enabled epistemological freedom in comparison, for instance, to sociology or psychology, where clearly very different perspectives have led to very severe conflicts. Also, maybe a common understanding that so-called primitive attitudes, such as naïve realism, are constructive, both in learning and research, contribute to a generally forgiving attitude when it comes to accepting such attitudes in spite of their philosophical difficulties. To think that you really deal with reality in your science – yes, fight with reality to squeeze the truth out of it – may help you endure the tedious debugging of your experiments or computer codes or the endless observing nights. Another possibility to explain this paradox may be the existence of a common understanding among us, that epistemology has only a minor significance in physics and astronomy. This would then be a remarkable attitude in sciences which have required fundamental philosophical re-orientations as a result of the scientific development of quantum physics and General Relativity. Still another hypothesis may be that big sciences, like physics and astronomy, require such large resources from society that an official facade of consensus among scientists is therefore important to demonstrate. There may be no place for doubts concerning the quality of science as judged by peer reviews when the costs for the taxpayers are impressive. Here again, a comparison with behavioural sciences is interesting. In many countries they have lost prestige and resources simultaneously with the playing out of dramatic conflicts about their fundamental epistemology.

Whichever of these explanations be true, we physicists and astronomers live in a happy world where the divergence in basic understanding is fortunate rather than problematic. In particular, we should appreciate the great tension between the Platonic tradition in theoretical physics and the rather brutal empiricism among the experimental physicists. That seems to be a very constructive difference which keeps science vital. What interests us are, in the end, methods that do work, that do produce interesting findings. The philosophically important distinctions, between, say, pragmatism and idealism, or between provisional truth and absolute truth, do create interesting discussions, but a hypothetical clarification of the real nature of scientific truth would, perhaps, not contribute much to physics. So maybe, it does not really matter. At least not to us.

Panel and General Discussion

Paolo de Bernardis:

Professor Gustafsson has selected the case and the results of COBE FIRAS as a proof of the Big Bang theory. So, is the Big Bang truth? Or is it provisionally true? What is truth in cosmology? This is the question I would like to ask and discuss. Let me say a few words about this and further developments we had in observational cosmology after these COBE results.

In a way similar to what physicists do, we cosmologists think that we can approach truth by measurement, and we think that what can be measured by different observers producing the same result within the measurement error, is true. However, as in physics, we know that, for a number of reasons, our measurements are not infinitely precise, and we need to take this into account. The work of the observational cosmologist is to invent methods and devices to improve the precision of cosmological observations, which, in general, are less precise than observations and measurements in physics.

In the last century, cosmologists have accumulated evidence of many facts important for our understanding of the universe – the existence of galaxies, the phenomenon of redshift, the distribution of galaxies and the initial abundance of elements, the cosmic microwave background, its spectrum, its polarization, its anisotropy – all these are measured quantities, so that we think they are true in the sense I just explained. But this is only the starting point for the construction of a science, and from these measurements a unifying theory has to be constructed to qualify cosmology as a science. Today, the most popular theory is the Big Bang, stating that the universe evolved from a hot, dense state and is still expanding and cooling from that initial state. According to a large fraction of the scientific community this is a solid theory, but there are notable exceptions. Indeed, according to another large fraction of the scientific community, there are grey areas, and there are details in the Big Bang theory which are not explained. There are undeniable and well-documented facts which are not explained by this theory. For all these reasons, I think it's fair to admit that we don't have a completely settled physical cosmology yet; and this admission is needed – first, because we need to be honest, and second because it stimulates further investigation, aimed at either validating or falsifying the Big Bang theory.

Which are these grey areas? Well, most evidently, the very beginning. We observe directly the light which arrives on earth. It is very old when it arrives here; it comes from 14 billion light years away. However, this was not the beginning: Indeed, the universe was already 400,000 years old when that light first originated in space. We don't have direct access to these first 400,000 years because, during that epoch, the universe was opaque, like the sun is internally opaque to light.

Another problem is the extreme isotropy of the cosmic background. This is difficult to explain and involves the development of yet another theory, inflation theory, which, as an addendum to the Big Bang theory, establishes a link between subatomic and cosmological dimensions. However, while very appealing, inflation theory has not yet been demonstrated.

Then, the problem of how structures form in the universe – the fact that the dynamics of galaxies and clusters of galaxies are not well fitted unless we introduce the so-called 'dark matter' component, which helps structures to develop and helps in binding structures in the universe. But here too, we lack stringent proof.

And, finally, there is the problem of the age of the universe, and the problem of the apparent 'dimming' of high red-shift supernova explosions. This dimming is interpreted as proof of an acceleration of the expansion of the universe requiring some strange 'force' driving the acceleration. We call it 'dark energy'. It could be vacuum energy, or something like that, we really don't know what it is.

The problem is that all these different kinds of 'cosmological evidences' require different ad hoc explanations. However, this is contrary to our 'scientific likings': we like simple, symmetric theories; and the addition of such extra ad hoc elements is disturbing and, in some ways, casts doubts on the truth of the main corpus of the Big Bang theory.

My interpretation is that the main corpus of the theory, the fact that the universe is expanding from a very dense, hot initial state, is solid. This is confirmed by a number of observations. By contrast, these unproven 'evidences' are not. Personally, I believe that only new experiments can solve these issues.

Some years ago, we produced maps of the cosmic microwave background coming from the universe. These maps were confirmed by other experiments, and they are consistent with the concept of this strange universe made of dark matter and dark energy with an inflation phase. So they are consistent. But is this enough to say that this strange theory is true? Probably not. We need more experiments. What is reassuring is that, for example, sometime this or early next year, we shall start two very important new experiments. One is the Planck satellite measurement of the cosmic microwave background, which will map the microwave background with unprecedented precision, and the other is the Large Hadron Collider at CERN, which also will study the nature of matter at extremely high energies never seen before. Both promise to address issues like dark matter. I really hope that these two experiments, probably together with yet other additional experiments, will bring us closer to the truth in cosmology. This is my last word for the moment.

Bengt Gustafsson:

I would like to say that I share all the views you expressed. In a sense, I think we have this very hot, very dense state, but the details about it and how it came about, are unclear. I agree with you that we are, in a way, in much worse a situation than we used to be, because the large-scale phenomena are driven by two 'entities' which are unknown: the dynamics of the universe are almost totally determined by the dark matter, consisting of at least five times more matter than the observable matter, and we don't know what that is. And then we have the so-called 'dark energy', which serves just to accelerate the expansion of the universe and appears to correspond to 70% of the universe's total energy. So, there is a very small percentage of the universe, of its matter and its forces that we can observe directly, at least at the moment, and this is highly unsatisfactory. To make things worse, we know nothing about the beginning: we don't even know if there was a beginning at all, or what was before. Or, alternatively, do we have to advocate the concept of this enormous ensemble of an extraordinary number of individual universes and us living in one of them? Obviously, our possibilities to scientifically understand why the universe is as it is, except referring to our own existence, are extremely limited and unsatisfactory. This may seem very unappealing to many of us, but that's how things are. We are in a difficult situation in cosmology, but in a very interesting and stimulating one.

Question (Lord May):

In your excellent lecture, you made a distinction between physics, where you can do manipulative experiments and astronomy, where all you can do is to make comparisons among observations. This difference is often misunderstood. No lesser a person than Karl Popper failed to understand the power of appropriately controlled comparative observations, and this led him to dismiss evolutionary biology as outside the realm which he, in his particular self-absorption, considered to be science. Whereas, in fact, by appropriately controlled observation and by compiling relevant relationships you can draw valid conclusions even without manipulative experiments. Thus, even in your third limit of cosmology, where cosmologists have only the one universe to study and cannot modify it experimentally. Even there, as you just explained so clearly, we *can* look backward in time to learn things and reach valid conclusions. I just wanted to reemphasize this and remind us of Popper's profound philosophic misunderstanding which resulted from an inadequate appreciation of the usefulness of the tools employed, because we may be in a similar situation with respect to cosmology today.

Question (Manfred Eigen):

I'm going to add something to what Lord May taught us about evolution-biology, and to extend it to molecular biology. We know that the proteins came about by evolution. So we started to study evolution using molecular biology, and, meanwhile, evolutionary biotechnology has become a flowering scientific field. My question is as follows: in biology, on the basis of Watson and Crick's discovery, we neatly calculated how probable it is that man would result by chance of the three billion nucleotides in the genome. The answer is that such a chance cannot be expected. You need evolution to explain man on the basis of his nucleotides. Now what about the occurrence of life in our universe? It has been calculated by some that there may exist some 10 to the 300th possible universes. What are the probabilities, in this case, that life appeared by chance? In other terms: should we have an evolutionary theory in cosmology too? Can you construct such a theory on the basis of phase-transitions and periods of exponential or hyperbolic growth – of which we know that they existed?

Bengt Gustafsson:

I think many of us would be very happy, if one could construct an evolutionary theory which would make it kind of natural for the world to have come out of some initial chaos into conditions where things might have been sorted out, ultimately resulting in the world we have now. We are, of course, very far from anything like this yet, but it is a dream, at least, that maybe we can construct such a theory sooner or later. But let me also say that there are severe problems with the multi-universe model also in terms of observability, and that, if we need to elaborate extra theories which are exterior to the present world to explain why the universe is as it is, then we are probably overdoing the theory.

Question:

Just a follow-up to what Manfred Eigen said: in evolutionary biology, a crucial element of evolution is the struggle for survival. Is there an analogy in the hypothetical evolution of the universe?

Paolo de Bernardis:

Well, if I can try to answer, I'm sure that the meaning of the term 'evolution' for cosmology is different from that which is used for evolutionary biology. For example, the phenomenon of feedback is essentially lacking in the current view of cosmological evolution, while it is extremely important in biology. So sometimes we use the same term, but with different meanings. It is important for us to be aware of this.

Question (Jean-Pierre Changeux):

Could there be a struggle for stability among universes? Just a struggle for stability, for existence instead of the struggle for life. It would mean that some conditions of – whatever – are more stable than others, and are then selected through this evolution?

Question:

Just to follow on this discussion about biological versus cosmological evolution: one fundamental difference appears to be that in biological evolution we are talking about differential survival of self-replicating entities, and I assume one would not propose that the universes are self-replicating entities.

Question:

Coming back to physics, Professor Gustafsson, what do you think about particle physics, about quantum dynamics etc.? These are theories that work very well, but their mathematical foundations are considered consistent by some, and as more or less 'ad hoc' by others. So there is no mathematical beauty or mathematical soundness about them. We are working with theories which are accepted just because 'it works', and that seems very disappointing.

Bengt Gustafsson:

Well, I agree with that. At the same time, it is also a wonderful theory, of course, if you look at what it actually can accomplish. But I should say that I have been told that superstring theory may be able to remedy some of these persisting problems in the standard theory. Still, it's a very complex mathematical theory. But let's hope.

Session III: Biological Sciences – Biodiversity – Sustainability

17 May 2008 – 8:30 a.m.

Chair:
Nicole Le Douarin

Speakers:
Jean-Pierre Changeux *Is the Brain the Organ of Truth?*
Robert McCredie May *Science as Organised Scepticism*

Panel:
Marco Baggiolini

Nicole Le Douarin:

This morning's session deals with the life sciences – biological sciences, biodiversity and sustainability. The first speaker is Professor Jean-Pierre Changeux, 2001 Balzan Prize-laureate in cognitive neuroscience.

Is the Brain the Organ of Truth?

Jean-Pierre Changeux:

I feel honoured and pleased to be with you today to talk about the brain in the context of a symposium dedicated to the difficult question of 'truth'. Yet I may say from the start that I have to be humble, since as a neuroscientist I am not in a position to answer the question "Is the brain the organ of truth?" on the basis of irrefutable empirical data. I would prefer instead to suggest a *research programme* for the next decades on this fundamental topic. The philosophers and theologians who are present will thus be disappointed by the absence of satisfactory demonstration and my scientific colleagues will be disappointed as well by the limited number of reliable data. Nevertheless, I will just take the risk to carefully explore with you what might plausibly be done, both theoretically and experimentally, in the next decades about the relationships between brain and truth.

Balzan (ed.), *Truth in Science, the Humanities and Religion: Balzan Symposium 2008*,
DOI 10.1007/978-1-4020-9896-3_4, © Springer Science+Business Media B.V. 2010

I will start by showing a painting of Charles de La Fosse, *Christ in the House of Mary and Martha*, which my wife and I donated to the museum of Meaux. In my opinion, it may be useful to illustrate Ricœur's concept of the "capable person." Martha, represented on the upper left of the painting, is asking, "Lord, don't you care that my sister has left me to do the work by myself? Tell her to help me!" Then Jesus responds, pointing his finger to Mary sitting at his feet, "Martha, Martha, you are worried and upset about many things, but only one thing is needed. Mary has chosen what is better, and it will not be taken away from her." I'm not going to discuss the rather ambiguous moral content of the answer, but just say that we are dealing here with *intentional communication* in a social group limited to three partners. I may suggest that, in the brains of the individuals who are represented in the picture, neural representations of the words and statements that are communicated are generated – of their meaning, the stored memory of the past work respectively accomplished by the two partners, and in particular, the neural implementation of self-recognition and of self/other distinction. There is evidently what Ricœur calls "a search for mutual recognition" between Mary and Martha, and what I may qualify here as a request for equitable shared rewards. We are therefore in the position to possibly examine, in neural terms, what a capable person might be, with her/his personal identity and social status in a defined individual, socio-cultural and historical context. My point is that, from the start, we have to avoid a simplistic and naïve view of brain's biology and consider the issue of brain and truth, and approach the question "Is the brain the organ of truth?" more specifically, within the complex and difficult framework of the capable person. Nevertheless, I cannot resist to quote from the start Spinoza's *Ethics* as a philosopher's response: "Men judge things according to the organization of their brain."

The Complexity of the Human Brain

The human brain is made up of billions of neurons and a million times more of synaptic contacts (around 10^5–10^6), which gives an enormous number of potential combinations within the brain. When I am asked the question "Do you believe that the *Last Judgement* of the Sistine Chapel or the *Sacre du Printemps* are actual productions of a human brain?" my answer is "Yes of course! What else could it be?" But, then the difficulties start.

Indeed, we have to deal with the brain as an extraordinary complex system with considerable plasticity that may be further considered, as discussed in my book *The Physiology of Truth*, as an open, self-organising, synthesis of multiple nested evolutions.

First, the human brain originates from that of the ancestors of modern man through the evolution of the species, and of their genome, within a timescale of millions of years. *Homo habilis* lived about three million years ago. And then, each individual brain develops, from the fertilized egg and the neurula up to the adolescent brain through neurogenesis and cell deaths, with a timescale of months and years. Finally, the synaptic connections between neurons become established through multiple outgrowth, stabilisation and elimination processes in the course of the postnatal development of the brain, which lasts years up to the adult stage. And still some synaptogenesis and even neurogenesis continues in some territories of the adult brain. The interaction with the environment controls this 'epigenetic' evolution, which gives rise to the production, storage and transmission of cultures. In addition, there is an internal evolution, through rapid changes of synaptic efficacies in the 100 ms to s timescale, that results not only in the production of mental objects and the dynamics of thought, but also in the intentional communication between brains and their evolution. As a consequence, social and cultural evolutions may take place with additional storage of information in extra-cerebral memories like inscriptions, books, works of art, and of course scientific publications.... These social and cultural evolutions are internalised within each individual brain, together with the personal history of each person. The brain is thus an extremely complex system, constantly evolving from the molecular level to the social/consciousness level, with nested hierarchical and parallel levels of organization constantly interacting with the outside world in a projective style. In addition, concomitant *bottom-up* versus *top-down* regulations integrate these multiple evolutions. In a more general manner, the human brain has to be viewed as a social brain, as 'an intentional organ', which is a spontaneously active, open, motivated and self-organizing system with built-in access to consciousness and shared social recognition. As a consequence, we may adopt, together with Bachelard, the position that neuroscience, like chemistry, is grounded in a sort of "educated materialism".

What Is True Knowledge?

Many definitions of "truth" have been presented this meeting. Yet several of these definitions may not necessarily be contradictory to one another. For instance, for Baruch Spinoza, "a true idea must agree with the object of which it is the idea," a definition that fits in with Denis Diderot's "conformity of our judgments with things." For Bertrand Russell, "beliefs are true when congruence obtains between object and belief" when "an isomorphism [exists] between the structure of thought and reality." The position of René Descartes differs: according to him truth arises as "the evidence of our reason," just as it does for Immanuel Kant, for whom truth is "the agreement of knowledge with itself with the abstraction of all the objects and all the differences between them." Finally, for William James, "truth is in fact an event, a process… a process of validation."

I will divide my presentation into several sections that may lead to the suggestion of neuronal models that may fit in with some features of these definitions: first, human *innate knowledge* associated with the universality of the *homo sapiens* genome; second, the postnatal *epigenetic selection of meanings* through cognitive games and testing truth with the external world; third, *consciousness*, the universal organization of thought and testing truth against the "inner world"; fourth, social communication, cultural diversity and mutual understanding, or beliefs and testing truth at the level of *the social group*; finally, the universalization of knowledge, the emergence of *scientific enquiry* and the issue of objective truth.

The Genetic Universality of the *homo sapiens* Brain and Universal Innate Knowledge

The genetic identity of the *homo sapiens* brain faces an important paradox. Let's have a look at the genome size or the number of structural genes, from yeast to worms, from flies to mouse, and then from mouse to man. There is a significant increase in the genome size and number of genes from yeast to mouse that fits in with the increased complexity of the organism, and in particular that of its brain. But when comparing the genomes of the mouse, chimp and human (for which complete sequences are known or to be known soon), one finds nearly the same number of genes (perhaps a slight decrease in the number of active genes from mice to humans) and a close similarity of the sequences. Yet, both the number of neurons and the complexity of brain synaptic organisation increase in a dramatic manner. How to account for this 'non-linearity' paradox? The answer is not to be found in any kind of mysterious force, but rather in the developmental genetics of the brain, which is indeed an important topic for research in the next decades. It lies in the patterning of embryonic gene expression, starting at the level of the early neurula. It may operate, for instance, through the spatio-temporal expression of transcription factors that regulate the differential expansion of the cerebral cortex and the morphogenesis of its diverse areas. Particularly critical in human brain evolution are the differentiation of the language areas and the relative expansion of the pre-frontal cortex from the squirrel monkey to the chimp up to man. One may hypothesize that these evolutions in patterning involve, among other things, process shifts of boundaries caused by minor genetic changes in regulatory loops established in a combinatorial manner between transcription factors. Plausible models of these mechanisms have been elaborated but not yet demonstrated.

Still, one may say that the evolution of these networks of gene interactions plausibly account for the innate functional organisation of the human brain and thus the universal innate knowledge of the human newborn. Among such forms of innate knowledge, one may quote: breast sucking, the recognition of the human face, the distinction between living organisms and inanimate objects, primary consciousness, predisposition to empathy and sympathy, and so on and so forth. As a consequence of the genetic evolution of the human brain, a first *true adequation* of brain functions with the outside world, which is necessary for the survival of the species, becomes established.

The Epigenetic Selection of Meanings Through Cognitive Games and Testing Truth with the External World

The brain of the newborn human strikingly differs from that of the adult. Looking at the development of the infant brain from birth to adult, the weight of the newborn brain – this is just an index – increases fivefold. In the course of postnatal development, the dendritic arborisations of the pyramidal cells in the cerebral cortex and their spine number increase considerably. About 50% of the adult synapses (which is a huge number) are established after birth in the brain. In 1973–1976 with Courrège and Danchin, I proposed a model for the developmental *epigenesis* of neuronal networks by selective stabilisation of synapses. The model suggests a non-Lamarckian mechanism for how the environment lays down traces in the developing brain through synapse selection. First, a genetic envelope specifies how the main patterns of connection are endogenously established between developing axons and their targets, yet not in a strictly one-to-one relationship. At a sensitive or critical period of development, some kind of exuberance of the synaptic connections occurs, a transient variability of the connectivity takes place. Then processes of elimination and/or selective stabilisation occur under the control of the state of activity of the developing network. Since Wiesel and Hubel's pioneering experiments (1963) on the development of the visual cortex in the cat and monkey, strong supporting evidence in favour of this scheme has accumulated, with many demonstrations of synapse elimination, or 'pruning', during development. To the extent that one may say, as I wrote in *Neuronal Man* in 1983, "to learn is to eliminate". These processes of selective stabilization/elimination of synapses through interaction with the social, physical and cultural environment are of critical importance to understand the development of the human adult brain. They may, of course, bring additional answers to the abovementioned non-linearity paradox.

Before examining with you how synapse selection may take place, I would like to illustrate how representations of objects from the outside world may actually 'enter' our brain. Using an auto-radiographic method, Tootell and colleagues were able to reveal two-dimensional patterns of activity elicited by the controlled stimulation of the retina in the primary visual cortex of the macaque. Using, for instance, a stimulus that has the shape of a wheel with spokes, one can recognize a similar figure, though with some deformations, in the cerebral cortex. The same result is achieved with a set of parallel bands. At the level of the primary visual cortex, some kind of isomorphism – remember Bertrand Russell's words – can be recognized between the stimulus and the neural representation. Yet such relationship is being progressively lost moving up to the higher visual areas and the prefrontal cortex.

Concerning how the selection of defined patterns of connections may take place when a representation has entered the human brain, one of several possibilities that has been suggested is the mobilisation of the midbrain reward systems (Gisiger, Kerszberg and Changeux 2005). Among the neurons involved in reward, one may mention the dopaminergic neurons from the midbrain and the opponent serotoninergic neurons. Accordingly, a positive reward received from the environment may result in the selection of the particular pattern of neurons that elicited the rewarded response. This is an efficient way to establish an *adequacy* between an internal brain representation and the outside world. This theoretical proposal has been simulated on the computer to illustrate the synapse elimination-selection process that takes place in the course of the storage of a representation.

This is just to show how the selection of representations may happen in the brain through the interaction of the newborn with the outside world. Comparison of the development of the human brain with that of the rat or of the monkey shows, in addition, that the period of synaptogenesis is much longer in humans than in the cat or the macaque. This is of course inscribed in our genetic endowment, but again, it may not 'cost' a large number of genes, since it deals only with an extension of the period of synaptogenesis outside the uterus of the mother. Moreover, one notes that the envelope of the total number of connections shows a significant decrease before puberty occurs, a massive pruning that is not compensated for by the concomitant increase of the total number of synapses taking place through development. Interestingly, this long postnatal evolution coincides with the internalisation of the socio-cultural environment of the newborn. One might even say that the epigenetic development of the brain is at the origin of culture, its diversity and its evolution.

Reading or writing are certainly not inscribed in our genetic equipment since writing is a pure cultural discovery that took place about 5,000 years ago in Mesopotamia. There are nevertheless unambiguous neural circuits involved in reading and writing which can be altered, for instance, by lesions consecutive to vascular accidents. At the turn of the twentieth century, Dejerine had already noted cases of pure alexia (i.e., loss of reading ability) without agraphia (or deficit in the processing of writing), indicating the occurrence of specific neural pathways for reading and writing. Brain imaging studies by Ingvar and Castro-Caldas reveal further striking differences in the gross organisation of the brains of literate and illiterate adults. Reading and writing would then result from an epigenetic appropriation of developing circuits. *Cultural pathways* are thus established in our brain in the course of its postnatal development.

Wittgenstein developed the concept of language games for language acquisition. I may take the liberty to extend Wittgenstein's view to knowledge acquisition in general, and propose the notion of *cognitive games* for the selection of representations that match versus mismatch the outside world and lead to the acquisition of basic semantics in the child as well as of symbolic systems and moral rules. I will return to this issue.

In the course of postnatal development, a postnatal epigenetic selection of meanings through cognitive games takes place together with their *testing for truth with the external world.* An epigenetic 'adequation' of the brain with the outside world further develops on top of its innate dispositions, thus making plausible Denis Diderot's definition of truth as "a conformity of our judgments with things." This does not necessarily mean a precise isomorphism "between the structure of thought and reality," as suggested by Bertrand Russell.

Consciousness, the Universal Organization of Thought and Testing Truth Against the 'Inner World'

In recent years, the development of theories and technologies and the accumulation of empirical data have legitimated a scientific approach to the neural bases of consciousness. First, I might try to suggest a definition of consciousness. One may view consciousness as a 'subjective space', an *internal physical milieu*, where actions are replaced by simulations, plans, goals and courses of actions, which are processed and evaluated in a *global* manner – I emphasize the word 'global' – and where mental syntheses and decisions occur with reference *altogether* to the outside world, the self and personal memories, internalised rules and social convention. Another important aspect of conscious experience is that one may report it. You may say, for instance, at tea time, "I listened to Changeux's talk this morning and he said this and that." When you make such a statement, you consciously recall a past conscious experience that simultaneously involved several perceptions, e.g., visual and auditory, as well as an understanding of multiple meanings. One might say that these events that have been stored in the brain and consciously reported are neuronal processes realized in brain structures.

Moreover, to avoid confusion, I think it is appropriate to make the distinction between what is referred to as the *states* and as the *content* of consciousness. The states, or levels, of consciousness include sleep, wakefulness, dream coma, and epilepsy, and they are under the control of neurons from the brain stem that release neuromodulatory transmitters like acetylcholine and/or norepinephrine. The content of conscious experience is what you can report about a past experience. Here, we are concerned exclusively with this second aspect.

The model of the global neuronal workspace (GNW) that I shall now discuss was initially designed by Stanislas Dehaene and myself (1998) to account for the 'global' and multi-modal character of the conscious space. It distinguishes two main computational spaces: *specialized processors*, which are modular, encapsulated, and automatic, and include perceptual systems, long-term memory (including autobiographic memory and self), and attentional and evaluative systems; and second, a *global neuronal workspace* where mental syntheses and decisions are processed and that broadcasts signals to multiple areas of the brain, yielding subjective experience and reportability. The critical anatomical hypothesis proposed is that the global workspace consists of a distributed set of cortical neurons characterized by their ability to receive horizontal projections through long-range excitatory axons from homologous neurons in other cortical areas and send them back again. Such long-range corticocortical tangential connections are mainly known to originate from the pyramidal cells of layers 2 and 3. Moreover, these neurons from cortical layers 2 and 3 are most abundant in the prefrontal cortex, and establish a long range reciprocal circuit with the parietal, temporal and cingulate areas. This global network of long-range axons would give access to the subjective experience of being conscious and, of course, to reportability since it interconnects multiple territories that include the language areas.

The main virtue of this still hypothetical GNW model is that it has stimulated a number of experiments, in particular, on the access to consciousness – for instance of visual stimuli. Conditions may be designed by rapidly flashing sequences of words and masks where a given word is either consciously perceived (the subject reports the word) or non-consciously processed (the subject does not report any word but is nevertheless primed by the word.) Brain imaging studies of the subject carried out under these two conditions reveal that in the course of conscious processing, the prefrontal cortex in particular is activated, while in the case of the masked words it is not (Dehaene and Naccache). A natural history, I might even say a biophysics, of consciousness can thus be objectively envisioned in humans. Recent progress has shown that the dynamics of the access to consciousness has been assessed by electro-encephalography revealing time ranges for access to consciousness in the 150–200 ms.

Tests for logical reasoning, which may be viewed as some kind of conscious internal tests for truth, have also been developed in humans and examined by brain imaging (Houdé). Again, an activation of the pre-frontal parieto-temporal cingular network is detected. The self-evaluation of a conscious melody that takes place in the course of reasoning, as I might call it, thus plausibly mobilizes the GNW circuits. Hopefully, we might some day be in a position to implement, in neural terms, René Descartes' statement that truth arises as "the evidence of our reason"!

Social Communication, Cultural Diversity and Mutual Understanding: 'Beliefs' and Testing Truth at the Level of the Social Group

As already stated by Aristotle, the "human animal" is altogether rational and social. An important theme of today's neuroscience is the neural bases of communication between human individuals in the social group. An interesting paradox is raised by the fact that, despite the considerable epigenetic variability existing between individual brains, members of the social group can understand each other. In the case of oral communication, a variety of sounds are used that are linked to meanings. Thus, a 'normalisation' has to take place *between* individuals, so that for each individual the same sound matches the same meaning. A model for such normalisation can be imagined on the basis of "shared rewards" (Changeux 2005). To illustrate this model, I will use a famous painting by Titian in The Louvre, *The Madonna of the Rabbit*. The painting represents St Catherine and the Holy Virgin Mary looking at Jesus Christ, the Christ Child, while He is looking at a white rabbit. You may now imagine that at this precise moment – sorry the sound is missing! – the Mother is saying "white rabbit" to her Child. In the baby's brain this establishes a neural connection between the sight of the rabbit and the sound which is produced by the mother and validated by the shared reward of the emotionally secure communication between the three partners. One may further anticipate that such shared rewards contribute to the internalisation of language in the newborn human brain through the above-mentioned cognitive games. Representation of the outside world and of ourselves – De Saussure "signifiés" – are thus linked with the arbitrary sounds or "signifiants" that compose the language of a defined cultural and social community. The shared reward between the two individuals causes the development of a communication system between individual brains in the social group, and the words they use will serve to test the 'truth' of representations communicated between them.

Claude Lévi-Strauss in *La Pensée Sauvage* has carefully studied the complex systems of myths, rituals, religious symbols and social practices existing in diverse societies and described their regularities and differences. He noted that most of the time these symbolic systems – or 'beliefs' – are in contradiction with elementary physics and common sense, but nevertheless are accepted as true and stably perpetuated from generation to generation in a given social group. Moreover, the members of the group do not confuse these 'symbolic representations' with current empirical knowledge about daily life. But how and why are these symbolic systems conserved and unique to the human race? Here, we are facing an important issue in social life and communication. The French sociologist Emile Durkheim has emphasized the importance of these symbolic representations for the stability and identity of the social groups, and thus for the reinforcement of the social bond through shared rewards. In my opinion, such symbolic systems have been created by humans to bring 'imaginary' answers to questions that they spontaneously raised because of the advanced cognitive capacities of their brain: i.e., the origin of the world and its meaning, the rationale of death, the possibility of life after death...many of them of course still pending!

The Universalization of Knowledge, the Emergence of Scientific Enquiry and the Issue of Objective Truth

The epigenetic plasticity of the human brain has permitted – I would even say favoured – a broad diversification of these systems of beliefs according to history and geography, thus comparing humans from different traditions in terms of the veracity of these beliefs. My late colleague Jean-Pierre Vernant discussed how the ancient Greeks attempted to solve this difficulty. They initiated public debates that occurred between citizens in a common place of the City: the *agora*. The coexistence of multiple schools of thought and diversity of opinions was accepted, but they had to be publicly presented in the *agora*. Criticism ensued, and the solution that worked the best was then selected after harsh debates. The proposition that was best suited to the real world was accepted, beyond social conventions and particular symbolic systems. But it was constantly challenged and eventually revised. Is that not the best definition of what we define as scientific enquiry today: the search for objective, universal knowledge? The *agora* may thus be viewed as being at the origin of modern science's quest for truth!

One of the most important aspects of the quest for objective truth manifested by scientific research is that it will never end, simply because the representations arising from our brain are not expected to be exhaustive of reality. On the other hand, scientific knowledge about the world and about ourselves is evidently in constant progress. An ultimate validation of scientific knowledge can, in my opinion, be found in its applications to technology and its practical applications. This is particularly striking for medicine. The enormous progress achieved by medicine in the past centuries, is a real criterion of the validity of our understanding of human diseases and therapies. It occurred through the casual and erratic development of scientific theories and their experimental validation/invalidation. Thus, I find it legitimate to mention William James's statement, "Truth is in fact an event, a process… a process of validation".

Following the recommendation allegedly inscribed on Apollo's temple at Delphi and also closely associated with Socrates, "γνῶθι σεαυτόν" ("Know thyself"), a good incentive for our search for objective truth might first be to better know how our brain works.

General Discussion

Nicole Le Douarin:

Thank you very much, Professor Changeux, for this beautiful and so beautifully illustrated talk.

Question:

I'm not an expert in the matter under discussion and would therefore like you to explain why you have used the central term of reward and rewarding mechanisms in brain development. Could you expand on this? Reward seems to be extremely important in adult neuronal structures, and here, a vicious circle may develop if we start talking about reward with all its historical complexities. Indeed, your contribution seems to give life to the phantom of Freud's pleasure principle and the principle of authority.

Jean Pierre Changeux:

Thank you very much indeed for asking this question. I am trying to be a neuroscientist. Freud was a neuroscientist at the beginning of his life, but unfortunately he did not continue along these lines. The question you ask is very important; it concerns the notion 'reward system', and I'm sorry to have used this term. It is commonly used by neuroscientists working on the dopaminergic system. Maybe, I should have employed the term 'emotions' instead, which would perhaps be more appropriate and where, as you know, more complex systems are involved. And of course the issue is how these emotional reward systems have been selected through evolution. They have been selected indeed. And they have been selected through evolution to save experience to the human individual. Because if the subject were to actually go through all experiments which are bad for his survival, he would not survive. Indeed, in the course of evolution, not only has there been acquisition of this innate knowledge, but also the selection of a system of evaluation, which is such that you are enabled to have judgments and opinions about what is dangerous for you or not, e.g., through the conscious workspace within our brain. So these reward systems have been preselected, but they are also subject to learning. And when I talk about the internalization of ethical rules during the development of the newborn, I wish to say that there is an interaction between an innate moral endowment – the basic principles of some kind of universal innate rules of social regulation, which are at the origin of the neural basis of ethics – but which will then be regulated and modulated through acquired knowledge, in order to adapt and fit this innate disposition to ethical behaviour through the cultural environment in which you live. And in my opinion many of these basic rules for ethical behaviour have been utilized by the different cultures and embedded into books or rules or whatever. In fact, you may find that they are very similar throughout numerous (if not all) human cultures. Because, if you take the Golden Rule, for instance, you find it in Confucius, you find it in the Old Testament with Hillel, and you find it with Paul also, and many others. In my opinion, this means that there are universal rules which are linked to the organization of the human brain, and they can be fitted and adapted to a given culture and social environment through learning.

Nicole Le Douarin:

I now ask Lord May, 1998 Balzan Prize winner in biodiversity, to give his talk.

Lord May:

I begin with an apology. I am a typical Anglo, and I speak only one language. I envy and admire those who, like most of you, speak two or more languages, and I wish I could give this lecture in Italian, but you would not like me to try.

Science as Organised Scepticism

Robert McCredie May:

Under the general heading of 'Truth' – the theme for this meeting – I find it helpful to make a necessarily very rough distinction among three different kinds of truths or ways of seeking understanding. The first might be called *mathematical truths*, in which one starts with a set of defined axioms or propositions and then deduces the consequences within a closed logical system: if two sides of a Euclidian triangle are equal, then two angles are equal. Although I call this "mathematical truths", it includes a wider class of logical exercises, puzzles and paradoxes. The second I call *science*, embracing social sciences, engineering and medicine along with the more narrowly defined physical and biological sciences. Here we are trying to understand how the real world actually works, from the age of the universe to the molecular machinery whereby our planet's living things assemble themselves to the fundamental structure of elementary particles. All such understanding is contingent, achieved and progressively refined by testing ideas against experiments and empirical observations; by now some things are effectively certain (the second law of thermodynamics), whilst other important questions remain as yet unanswered (how to design a vaccine against HIV/AIDS or how to minimize systemic risk in financial systems). My third category addresses our quest – older than recorded history – to understand our own place in the universe and its meaning: our *beliefs and values*, along with their origins (often in religions, usually supported by canonical texts) and their expression in legal and ethical systems and other codes of conduct and in the creative arts.

I strongly emphasize that this tripartite codification is primarily to help organise this essay, which will concentrate mainly on the first two: mathematical and scientific 'truths'. I will touch only briefly on the category of beliefs and values, recognising that my thoughts may not pass muster against the professional rigours of discourse in these millennia-old realms of philosophy.

In what follows, I will begin by briefly discussing each of the above three kinds of truth. This will be followed by a discussion of how the three categories overlap in various ways: the most obvious is that for mathematics and science. The concluding section sketches some general rules for handling the interplay among these "truths" when making policy decisions, with illustrations from climate change and other topics. As will be seen, these general rules are more easily stated than implemented.

Mathematical Truths

For this category, in the general sense defined above, we are exploring the ineluctable logical consequences of a given set of assumptions or axioms. For generations of schoolchildren, Euclidian geometry could be the canonical example, with individual students finding the proofs difficult or surprising or even enthralling. But *quod erat demonstrandum* means just what it says.

Of course, even here the reality can be more complex. For one thing, we can often be fairly sure a theorem is true, but unable to prove it. Over a century ago, Hilbert famously set out a list of important but unsolved mathematical problems. The Clay Foundation marked the turn of the Millennium by announcing million dollar prizes for proofs of the seven still-unsolved Hilbert problems (one of which has now almost certainly been solved). Some of these problems, moreover, have important practical applications: a solution to the Riemann Hypothesis, for example, would have serious implications for widely-used methods of encrypting information. Fermat's "last theorem" resisted proof for centuries.

Particularly interesting is the "four colour problem", which conjectures that no more than four colours are needed unambiguously to distinguish the territories in any conceivable map. Having frustrated mathematicians for centuries, this theorem was relatively recently proved by a computer programme which explored a vast number of categories of 'exceptional cases', thus eliminating all possible counterexamples. Many feel that such an ugly proof, which provides essentially no deeper insight into this mathematical problem, is undeserving of the appellation 'proof'. A similarly brute-force computer proof has very recently shown that, as long surmised, the game of draughts or chequers is a draw with best play on both sides. One day, we may even have a similar proof for chess (for which the question is, in principle, clear), but here the corresponding computation is so huge that any such result lies in the far, far distant future.

Other relatively recent branches of mathematics deal with the 'truths' inherent in identifying and quantifying uncertainty. Mathematical statistics gives us the relative probabilities of a range of outcomes, given specific assumptions about the system in question (here the origins lie in gambling with cards or dice). Most recently, chaos theory lays bare the mechanisms whereby simple and completely deterministic rules or equations can, if sufficiently nonlinear, produce outcomes so sensitive to initial conditions as to be unpredictable (the very simple equation $x(t) = a\,x(t)\,[1 - x(t)\,]$, with $4 > a > 3.57..$ is one example). Looking ahead to the next heading – scientific truth – this undercuts the Newtonian dream that if you know the underlying physical laws, and they are sufficiently simple, then you can predict the future; chaos theory shows that even the simplest Newtonian clockwork can be unpredictable. And for the interaction between mathematical truths and beliefs/values, some have (perhaps playfully) suggested that chaos theory reconciles the foreordained Calvinist world with our illusion of free will!

A more radical issue is raised by Gödel's Theorem, which shows that some kinds of mathematical questions are simply undecidable. The "class of all classes not members of a given class" is perhaps the simplest example, although some would prefer to sequester such problems, relabelling them as paradoxes.

Scientific Truths

The attribute which primarily distinguishes *homo sapiens* from the wonderful diversity of other creatures we share our planet with, is our self-conscious quest to understand the world around us and our place in it. The first stirrings of this search are lost in mists of magic and mysticism, whose enigmatic traces remain in the caves at Altamira, in stone circles, and in other such reminders. Both archaeological and anthropological evidence suggest that this early quest involved a blend of my headings of Scientific Truths and Beliefs/Values. The scientific part was observational and descriptive: movements of the stars and seasons; the behaviour of other animals; the nutritional and medicinal properties of plants. This represents a very important beginning. But for most of history, the search for underlying causes or laws comes under the Beliefs/Values heading.

Going beyond description and observation, real progress in understanding how the natural world works only truly began with the Enlightenment, with its guiding principle that the truth is to be found not by appeal to authority but by experimental tests and evidence. Established in 1660, the Royal Society's motto embodies this precept: *Nullius in Verba* (which roughly translates as "Look at the evidence").

In school, in university, and especially on quiz shows, we tend to see science as certainty. Such a false impression is both understandable and unfortunate. Understandable because the science taught in schools, and in much tertiary education also, is about things we really understand very fully. This, after all, is a sensible way to organise curricula. Even more, the answers to 'science' questions on TV quiz shows cannot admit debate, and so help promote the illusion that science is a mass of boring trivia about the technical names for things, or definitions memorised by rote. Although understandable, this is deeply unfortunate. Whilst much of science deals with things that are indeed extremely well understood, many of the topics that engage public attention lie at or beyond the frontiers of the currently known. In fact, science is best seen as organised scepticism. The result, as we have seen for example with climate change, is a journey, over time, from uncertainty toward understanding, guided by experimental tests and sceptical questioning.

To put it another way, on any scientific issue, there is, at any one time, what might be thought of as a *landscape* of opinion. In the early stages of research, various ideas are proposed, producing lots of little clusters of opinion like little hillocks on the landscape. As questions are raised, observations noted, and experiments designed to discriminate among possibilities, some of the bumps shrink to zero, others attract support and grow. As things progress, we tend sometimes to find two 'schools' or hilltops, rarely of comparable size, or perhaps many hilltops, with some growing and others fading. Surprises are always possible. A single crucial observation may destroy an 'agreed' peak, and eventually create a new one – as the Michelson-Morley experiment did to the ether. More usually, as for special relativity in relation to Newtonian physics, a revolutionary transformation of a mature disciple adds important features to parts of the mountain, leaving most of it effectively the same (so that old maps still work in almost all circumstances).

Ideally, over time, confident understanding emerges, and appears as one tall, triumphant peak on an otherwise vacant landscape. In practice, things are rarely this simple. Pockets of aberrant opinion may hold out, with proponents either ignoring decisive evidence and experiments, or alternatively inventing ever more baroque ways of modifying their views to accommodate such facts. And, of course, such situations arise even more implacably if some fixed ideological position commits an individual to a position, transcending any data.

In general, however, the historical landscape of any major scientific question has its own evolution, from pimply plurality, through contending hillocks, to a single and narrowly-spikey mountain in maturity. It is important – although often difficult – at each stage in the evolution of such a landscape to maintain a clear sense of its geomorphology.

Unfortunately, the media's praiseworthy aim of always presenting a 'balanced' account can have difficulty tracking such an evolving landscape. The temptation, whether in print, radio or TV, is to seek to present the 'two sides', as if reporting a sporting event.

A particularly clear and topical example of this is to be found in the recent debates about the extent to which human activities – burning fossil fuels and putting greenhouse gases into the atmosphere – are causing global warming, as distinct from warming deriving from natural causes. Early warnings were suggested over a century ago. But we are dealing with a hugely complicated and nonlinear dynamical system, and even three decades ago there were very significant uncertainties associated with our quest for understanding. With computational power doubling every 18 months, things have greatly advanced since then, and the basic facts are now unambiguous enough to be agreed (although there do remain significant uncertainties about the speed of some processes – ice caps melting, or permafrost thawing, or oceans becoming more acidic, and so on). The basic facts are now clear enough that in 2005 the G8 science academies along with China, India and Brazil asked that we

> identify cost-effective steps that can be taken now to contribute to substantial and long-term reduction in net global greenhouse gas emissions [*and to*] recognise that delayed action will increase the risk of adverse environmental effects and will likely incur a greater cost.

These academies created the process of organised scepticism which we call 'science', and their recognition that we are well past the point where we should doubt the serious consequences of climate change deserves attention. But this virtual unanimity within the community of relevant scientists is bizarrely at variance with a recent opinion poll in the UK, which found 56% tending to agree (11% of them strongly agreeing) that "many leading experts still question if human activity is contributing to climate change"!

In short, science is organised scepticism: as much a way of knowing, a collective endeavour of asking questions, as a set of facts. Properly pursued, the process seeks experiments and observations that distinguish among alternative possibilities, progressively closing in on a better understanding of the phenomenon or problem under study: simplifying the topology of the scientific landscape until eventually one clear peak remains. Obviously this is an activity practised by fallible humans, individually and – most importantly – collectively. There are thus those who draw the conclusion that the product, no less than the process, is a social construct. This is a woeful misunderstanding. Yes, the choice of problems for study – the agenda and even sometimes choice of favoured programmes – can have elements of fashion and personalities woven into it and to that extent is, in part and only in part, a product of time and place. I think the same is often partly true for some of the ideas put forward in the very earliest – many tiny hillocks – stages of forming the landscape in a new area of science. But the knowledge secured as the scientific landscape, over time, narrows to a peak – the Second Law, the inverse square law, the fact that HIV causes AIDS, the Navier Stokes equations underpinning the computational fluid dynamics that designed the planes you fly in – can be absolute. As Max Planck wrote:

> there is a real world independent of our senses; the laws of nature were not invented by humans, but forced upon them by that natural world. They are an expression of a rational world order.

Beliefs and Values

Here I stray beyond my domain of competence. There are, of course, important questions which – while they eventually may make contact with our growing understanding of neurobiology and the evolution of behaviour of individuals and communities – currently lie largely outside the domain of science: questions such as, what is happiness; how do we handle the trade-offs between the rights of individuals and the collective interests of society (e.g. the arguments for and against compulsory childhood vaccination or identity cards); ultimately, what is the meaning of life?

For many, answers to essentially all such questions are ultimately provided by one or other of the world's religions, monotheistic or otherwise. Here the canonical texts set out the basic principles which have, in effect, usually been elaborated and interpreted by successive generations of learned teachers and scholars. Other belief systems, based for example on the writings of Marx or Mao, although disclaiming any religious connection, have served a similar purpose over the past century and more.

More generally, although the Enlightenment is characterised by its espousal of liberty of individual conscience and rejection of the authoritarian certitudes of dogma or revelation, the values it endorses build on the best of what went before. My personal belief is that most of the world's population either do, or would if they were allowed, embrace the values enshrined in the American Constitution: life, liberty, and the pursuit of happiness. These are also implicit in the United Nations Charter, and more recently in its Millennium Development Goals.

However, in a finite world under increasing pressure from too many people, with consequent difficulties for supplies of food and water, and for the survival of other species, there are inherent conflicts between actions that we need to take collectively, and the costs to individuals that these imply. This leads us to the central truth of our time, namely that we need better understanding of the evolution of cooperative behaviour – better understanding at the interface between scientific truths and beliefs-values.

Science and Beliefs/Values

From a 'scientific truth' point of view, how did cooperative associations among humans evolve, leading eventually to the complex societies of today? At first glance, the answer seems easy. You pay some small cost to gather a much larger cooperative benefit. For example, a prairie dog takes a personal risk in giving an alarm call, but all colony members benefit and, by taking turns as alarm giver, each individual's group benefit exceeds the occasional risk. But any such arrangement is immediately vulnerable to cheats who enjoy the benefits without paying the risk-taking dues. In evolutionary terms, such cheats have a selective advantage, and it is unclear how such observed cooperative phenomena can arise or be maintained. Darwin recognised this as his most important unsolved problem.

Following work on "kin selection" by Hamilton and others a century after Darwin, we now understand how such cooperative associations can evolve and be maintained in relatively small groups of sufficiently closely related individuals. These conditions could apply to humans when we were small bands of hunter-gatherers. But for large aggregations of essentially unrelated individuals, as developed once agriculture appeared and cities began, the origin of cooperative associations – with group benefits which exceed the 'cost of membership' – remains as puzzling today as it was for Darwin.

The problem has been much explored, in the scholarly literature, employing a variety of metaphors: the Tragedy of the Commons; the Free-Rider problem; the Prisoner's Dilemma; and others. The Tragedy of the Commons, for example, considers a 'commons', where all inhabitants of a village can graze their cattle. Left unregulated, each individual will add stock, until the commons is overgrazed. If some socially responsible individuals limit themselves, they will be economically disadvantaged in the short run, whilst still suffering the long-run degradation of the commons by their less inhibited neighbours. The analogy with climate change is obvious. This is a good metaphor, despite its historical inaccuracy. A more concrete and specific example is the recent decline in voluntary uptake of childhood measles-mumps-rubella (MMR) vaccination in the UK, as individual parents avoided a putative (but nonexistent, or at most undetectably tiny) risk, implicitly relying on others' children to be vaccinated to provide herd immunity; enough parents having made this choice, we are now seeing a rise in the incidence of measles (with a risk of serious consequences for roughly one case in 1,000).

Despite a literature running to thousands of papers on these metaphors in learned journals, along with much 'thick description' of real societies in many countries, I do not believe there is any agreed explanation for the evolution and maintenance of cooperative behaviour in human societies. Moreover – focussing narrowly on metaphors such as the Tragedy of the Commons or the Prisoner's Dilemma – despite the amount of published work there are essentially no studies exploring circumstances where costs and benefits vary among the players, as they clearly do in the real world of climate change.

My personal opinion – and this is pure speculation – goes as follows. Humans, as noted earlier, have probably always tried to make sense of the world around them. Once we move out of the mists of pre-history, we find stories of dreamtime, creation myths, ceremonies and initiation rites, spirits and gods, with a unifying theme that all seek simultaneously to help explain the external world and also to provide a 'stabilization matrix' for a cohesive society. There are, moreover, some striking and unexplained similarities in belief systems and rituals from different times and places. Conscience, a simple word for a complex concept which helps foster behaviour in accord with society's professed norms, has been memorably defined by H. L. Mencken as "the inner voice which warns us that somebody might be looking". And how helpful it is if that somebody is an all-seeing, all-knowing, supernatural entity.

Common to these conjectured 'stabilizing forces' in essentially all earlier societies are hierarchical structures, serving and interpreting the divine being or pantheon, along with unquestioning respect for authority. In such systems, faith trumps evidence. I think I was 8 years old when I first encountered, and was disturbed by, the biblical injunction, relating to the doubting St Thomas: "blessed is he that seeth not, yet believeth".

But if indeed this is broadly the explanation for how cooperative behaviour has evolved and been maintained in human society, it could be Bad News. Because though such authoritarian systems seem to be good at preserving social coherence and an orderly society, they are, by the same token, not good at adapting to change. Jared Diamond's book, *Collapse: How Societies Choose to Fail or Survive*, provides striking examples.

Deriving from the questioning spirit of the Enlightenment, our increasing understanding of physical, and more recently biological, systems have resulted in longer, healthier lives, liberated by energy subsidies and informed by easily accessible information about anything and everything. But these well-intended actions have also produced adverse unintended consequences. Many people and institutions have always found such questioning, attended by unavoidable uncertainties, less comfortable than the authoritarian certitudes of dogma or revelation.

I think this helps explain why fundamentalist forces are again on the march, West and East. Surveying this phenomenon, Debora MacKenzie has suggested that – in remarkably similar ways across countries and cultures – many people are scandalised by

> pluralism and tolerance of other faiths, non-traditional gender roles and sexual behaviour, reliance on human reason rather than divine revelation, and democracy, which grants power to people rather than God.

She adds that in the US evangelical Christians have successfully fostered a belief that science is anti-religious, and that a balance must be restored, citing a survey which found 37% of Americans (many of them not evangelicals) wanted Creationism taught in schools. Fundamentalist sects of Islam offer a complex but ultimately similar threat to science according to Sardar, Ruthven, Masood and others, who note that a rise in literalist religious thinking in the Islamic world in recent decades has seriously damaged science and free enquiry there, seeing the Koran as the font of all knowledge.

From this viewpoint, what we are seeing is not Samuel P. Huntington's "Clash of Civilizations", but rather a revival of an older clash between dogmatic, faith-based belief systems and the open-minded, experimental, questioning spirit of the Enlightenment. The really unfortunate thing is that none of these fundamentalist beliefs are grounded on, or representative of, the contemporary mainstream of the religions they profess to serve. Fundamentalist Christianity is widely considered as irrelevant to modern theology as it is to modern science. The extremist views and acts of fundamentalist Islam find little sanction in the Koran. Karen Armstrong suggests that

> to fight the secular enemy, fundamentalists reduce complex faiths to streamlined ideologies and, above all, try to recast old mythical tales as modern, literal truths.

In so doing, they tend to lose the compassion that is the mark of mature religious beliefs.

Mathematical and Scientific Truths

Here, little need be said. Mathematics is, in its essentials, no more – but no less – than a tool for thinking clearly. Thus a scientific idea about a particular phenomenon is often usefully explored with precision in mathematical terms. The mathematical relationship among facets of a problem can then lead to testable predictions.

In other contexts, the design of laboratory experiments, or the search for meaningful patterns in observational data, are facilitated by statistical techniques and analyses (which we often forget are, in their rigorous form, less than a century old).

In summary, although the certitudes of mathematical truths differ from the exploratory and organisedly sceptical nature of scientific truths, they work together in many ways. Indeed, one of the first applications of Newton's invention (simultaneously with Leibniz) of differential calculus, and arguably one of the motives for its invention, was to justify the replacing of a planet by a mathematical point of equal mass for the purpose of exploring orbits under gravitational attraction.

Mathematical Truths and Beliefs/Values

Less obviously, mathematics also intrudes – sometimes in very fundamental ways – into the implementation to deeply held beliefs and values.

For example, we can probably all agree that we should always elect the candidate with the greatest number of votes. Simple enough if there are only two candidates. But once we have three or more candidates (or other political choices among three or more options) we have a paradox first noted by Condorcet and by Dodgson (of *Alice in Wonderland* fame), and later independently elaborated in other contexts by Ken Arrow. With three choices, and all voters having logical 'transitive' preferences (if A is preferred to B, and B to C, then A is also preferred to C), the winner can depend on the voting system employed (or on the order in which pairwise comparisons are made). More generally, with N contestants for election, it is possible that there could be one winner on a single-vote basis, another on the basis of assigning preferences 1 through N to candidates and electing the highest scorer, yet another by allowing two votes (first and second preference) and successively eliminating lowest 'first preferences' and reassigning their second preference, and yet other winners using other methods! Admittedly the full panoply of these possibilities is unlikely to arise in practice, but the fact remains that the principle of 'elect the candidate with the most votes' allows considerable ambiguities in practice. And practical examples of Condorcet's Paradox have been documented.

Combining these remarks with those above about the evolution of cooperation, it is clear that truths associated with beliefs and values intertwine with seemingly separate ones associated with mathematics and science in ways which are many, varied and often highly relevant to today's problems.

Science and Society

Ultimately, as science advances our understanding of the external world, it offers us opportunities to improve life for all. But increasingly we recognise the unintended adverse consequences of well-intended actions, which threaten unprecedented problems, global in scale and scope, and ultimately deriving from the finiteness of our planet: climate change, shortages of water and other resources, new and re-emerging diseases, loss of biological diversity and ecosystem services, and more. As we address these problems, and as scientific understanding itself reaches down into the molecular machinery of life, it behoves us to think more carefully about which doors to open and which to leave closed.

In this task, the job of science is to frame the debate clearly, making plain the possible benefits and costs – and the concomitant uncertainties. And making clear that cloud-cuckooland is not a feasible choice. When it comes to acting out the democratic drama of choice on the stage thus set, science has no special voice; the drama of choice is about values and beliefs, about what kind of world we want.

All this is easier said than done. The difficulties in practice are many and varied. Many of those participating in what is advertised as a discussion of the scientific facts and uncertainties may in reality be bringing different agendas to the debate, like actors appearing on stage in disguise. A search for the objective truth, admitting attendant uncertainties – about climate change, or the evolutionary origin of species, or the safety of the MMR vaccine, or GM foods, or whatever – can too easily slide into something more like the adversarial games of the courtroom if some of the protagonists are committed to a set of beliefs or values which transcend any and all contrary scientific facts. Today, when clarity of fact-based understanding is increasingly important in addressing a range of environmental and social problems, such tensions between the organised scepticism of the scientific process and the comforting certitudes of dogma and belief are particularly unfortunate.

Panel and General Discussion

Nicole Le Douarin:

Thank you very much, Lord May, for this extremely interesting and thoughtful talk. And now I am asking Professor Baggiolini, who until very recently was the President of the University of Lugano, for his views on the presentations and discussions of this morning.

Marco Baggiolini:

Rather than commenting directly, I should like to point out one of the features of life sciences, experimental life sciences. This is a field where truth is atomized. We collect a lot of very little, very small, relatively easy-to-understand facts, and we almost never speak about truth. We generally speak about evidence, which is more or less the same. And I will spend some time to think about the system which has established itself to control this atomized evidence and to ascertain its veracity or truth, namely the large community of scientific peers. Indeed, I consider the opinions of peers one of the most efficient ways of validating experimental biosciences. And why? Out of interest: in such a broad field of science, many people belonging to the field will follow the literature and scientific presentations, and will have an opinion on what is presented. Many of them will be ready to correct what is presented and what, for them, is not as 'true' as they would wish. And this correction will be a contribution to a better understanding of the problems, and to the finding of truth, as in the metaphor Lord May was speaking about, where, among the mountains or hills, one suddenly rises to be a peak, visible to everybody. We have to realize that, in experimental sciences – in particular in biological sciences – truth may change daily. But ultimately the true, the right evidence will survive.

The capacity and the motivation of the peers to intervene are a consequence of the fact that they themselves profit scientifically from new ideas, from novelty and from new theories which emerge from facts. This is why peers are a community totally devoted to quality control.

In essence, the peer community plays two important roles: one role, which some scientists still fear, is the peer reviewing system, which helps reputed journals to improve the papers they publish. This system not only fosters scientific debate; it also contributes to the shaping of the scientists whose work is monitored in this way. At the beginning young scientists in particular may consider this an ordeal. However, with time, they not only learn to live with it, but they realize that they profit from it. The second is that of being able to actively 'vote' for or against new science (data, concepts, theories, opinions) by citing and quoting the relevant publications. And this 'voting' has become very effective with the appearance of citation analysis. I know that many scientists don't like to hear about citation analysis and it certainly isn't without flaws. However, it is not evil. It is a quick, fairly reasonable approximation of truth. This, of course, applies to experimental sciences; in other fields it is much more difficult. I stop here, and I hope that this was not too blasphemous a connection between Google and truth.

Question:

I would have questions to all three of you, but I shall start with Jean-Pierre Changeux: in one of your slides there was a definition of consciousness. Looking at it, I couldn't help noticing that all the components of that definition also apply to certain artificial systems. We have robots that are driven by an artificial neuron network, a little 'brain'; in the beginning their action is totally unstructured; then, based on a feedback-loop between the sensory input and the actions it executes, the system learns to do complex things like balancing a pole. It can even plan into the future a little bit. So there is planning, and apparently also a sense of self, because, to a certain extent, the system can evaluate the consequences of its own actions. So my question is: is that system conscious?

Jean-Pierre Changeux:

I have no difficulty predicting that, in the future, one might have robots fitting the description you give, i.e. having some rudiments of self-consciousness. I have no objection against that, and, in fact, one can envisage future technological developments in this direction. This has even led some colleagues to discuss the issue of the ethics of such robots. But we are still very far from this. However, I think the main difference between a robot and a human being, who, in a way, is a highly sophisticated robot, will always be the personal and individual history of any human being. With this, a human being reaches the status of what we call a person, which the robot is not going to attain. Maybe the closest it can come to it will be that perhaps one day one manages to build a robot capable of acquiring experience through the frequentation of human beings, but, here too, we are far from it. As I tried to illustrate, human beings have been selected through evolution to spend about half of their lives learning for their adult life. Because if you look at the early *homo sapiens*, his average lifespan was perhaps 30 or 40 years, and he spent at least 15 years to acquire knowledge for his adult life. This is amazing when you think about it, and I think that this is something no robot will be able to do, because, at the same time, he would have to store entries from cultural evolution, from cultural history and the evolution of science as well – if you want it to do mathematics or biology. At this stage, all these things are very difficult to imagine and conceive, but one should not exclude anything for the future. As I said, I don't see any contradiction between what I said and what you said and the present status. In brief, it's clear that we are very far from having even an approximate understanding of the processes of consciousness and, therefore, even further from conceiving consciousness in computers. I'm not going to say more than that.

Lord May:

Just a few words to observe that computers can do creative things, and my favourite example dates back almost 50 years, to the very early days of computers. In my talk I mentioned Euclid and proving theorems. I mentioned the isosceles triangle, with two equal sides and angles. It's one of the first things in the geometry text. This was given to an early computer to prove. Euclid's proof is to drop a perpendicular from the third angle to the unequal side. Then you see that the two sides are equal, the third side is common, the perpendicular forms a right angle, and therefore, the congruent angles have to be equal. The scientists wanted to test if the computer was smart enough to figure out it had to drop a perpendicular. No, it wasn't; it was better than Euclid. It said: "Triangle ABC is congruent with triangle ACB – the two having the 'common' side in common. Therefore, the four 'individual' sides of the two triangles being equal, the angles must be equal." This is a more beautiful proof than the two-millennia-old Euclid proof. So never underestimate what computers may do.

Question:

My question also involves computers, but does not put them in the centre. I would like to ask Professor Changeux whether, in his programme, there is already place for the quantitative, not just qualitative aspect of the work of the brain. There is one very interesting artificial situation where you can precisely say how much information the brain should handle in a given time, and compare it to a computer. It's the famous programme of the Deep Blue chess computer, which plays chess on the level of world champions. During about two hours, Deep Blue chooses each subsequent move in chess, and apparently does this at least as efficiently as a chess player of world class. My point is that we know very precisely what resources are required for a computer to use the Deep Blue programme, and that the brain apparently has no such resources. What strikes me, is the problem of time: one elementary choice by the computer takes some microseconds; but the speed of neuron actions is in the range of milliseconds, not microseconds. In addition, computers have a very efficient system of parallel work because they are very well synchronized, whereas in the brain synchronization is very crude. So, how can the brain match the computer? Do you think that, in the next ten or twenty years, we can hope to understand better such quantitative aspects of the work of the brain? Shouldn't this be an explicit question on the research agenda?

Jean-Pierre Changeux:

I think this is a very profound issue and question, and my answer will have three components. First, for the speed of the brain: I recently had a discussion with a mathematician about mathematics and the brain. He was amazed that the brain was not using electromagnetic waves like computers. He asked me, "Why does the brain not use the speed of light?" The answer is that our brain works with the mode of propagation of the nerve impulse we inherited from lower invertebrates through evolution. You know the action potential and its speed of propagation. The action potential was studied by Hodgkin and Huxley in squids, and in our brain we are using the same mechanism, elements of which – analogous to an acetylcholine-receptor though not in connection with transmission but with pH-sensitivity – have been discovered by one of my students in bacteria. Thus, already three billion years ago, there were analogues of the acetylcholine receptor we have in our brain. Our marine heritage of billions of years of biological evolution imposes constraints in our brain, and we have to live with it. So we cannot compare – this is the first answer to your question – the speed of a computer with the speed of the brain, the speed of the computer being many orders of magnitude faster.

Where is the solution? The solution lies in the organization of the brain and in how this organisation, through evolution, led to the capacity of differential quantitative versus qualitative evaluation of the world around us. One of my colleagues has been studying patients with lesions in the brain which create a dissociation between quantitative and qualitative evaluation. This is expressed for instance by a lady going to the market to buy a dozen of eggs. If she has such a lesion in the left side of the brain, she is unable to count twelve eggs, but nevertheless she is able to evaluate quantity, i.e. the global quantity that is present. When she sees the dozen of eggs, she is able to recognize it. So, we have ways in our brain to evaluate simultaneously qualitative and quantitative aspects, and, as you know, the computability, i.e. the ability for quantitative evaluation of our brain, is located in its left hemisphere together with the language areas, while the more global evaluation is in the right hemisphere.

Regarding our comparison with Deep Blue: Deep Blue is not using strategy like a chess player would. I am not a specialist in this, but apparently it is accepted that chess players learn chess playing like a language, and they learn to recognise, if I may say so, 'semantic situations' which are given by the pattern of objects in the field of the chess. If the brain recognizes a given pattern, then the players know which move they have to make. So in fact they have a very strong memory of diverse combinations of the pieces which they remember and learn. A good chess player is not using an extremely fast combinatorial analysis to identify the pattern, but has the pattern stored already and knows that, with a particular pattern, a particular move is good. So the chess players' speed results not from rapid calculations but from the fact that patterns are stored in his long-term memory.

Question:

I have a question for Lord May. I thank you for the 'tripartition' of truth – mathematical truth, scientific truth and ethical value truth. And I would like to apply it to stem cell research with embryos, which I consider to be at an interconnection between the first type of truth and the third type of truth. Indeed, the question is not whether human dignity begins at 12, 13 or 18 days, and I also don't know whether the question is whether the soul enters the embryo at or before that time, because I don't think that human dignity should be linked to the presence, the absence, or indeed, the existence of a soul.

And therefore we are left with the question of how to behave in the face of the uncertainty about the question whether the embryo is a human being or not. In the legal concept of precaution there is the principle of *in dubio pro securitate*, which implies that when in doubt one has to choose safety. In the case of human stem cell research versus human dignity this means that we should opt for the *in dubio pro securitate* principle and decide not to do research using human embryonic stem cells.

Lord May:

In view of the time, I make no attempt to answer that in detail. The point I was trying to make is that the church's formal position on abortion was quite different 150 years ago. It drew a clear distinction between abortion up to 40 days and after. I'm not quite sure on how it handled the question when it concerned a female foetus. Subsequently, the attitude changed as a result of an attempt to better understand science, given that the sacred texts were silent on the subject.

Question:

Just a short question for Professor Changeux. Do you care to comment on redundancy as an apparently useful property of biological systems, in particular in the process of evolution? Indeed, it seems that we are using only a fraction of our neurons effectively.

Jean-Pierre Changeux:

The thing is that there is indeed a transient stage of redundancy, or if you wish, of maximal diversity, in the course of development. In the case of the human brain you may have a small number of genetic modifications, which lead to the expansion of the size of the brain, of the relative surface of the cortex and thereby of the brains functional capacity. This will then be progressively 'filled' with specific 'functional content' through the interaction with the environment. So there is in fact a temporary increase of capacity with a reduction of redundancy afterward.

Session IV: History – Social Sciences

17 May 2008 – 10:45 a.m.

Chair:
Nicolette Mout

Speakers:
Quentin Skinner *Truth and Explanation in History*
Dominique Schnapper *Cultural Relativism: Meaning and Limits*

Panel:
Keith Thomas; Mauro Baranzini

Nicolette Mout:

History and social sciences are sisters – they sometimes quarrel, but not today, I hope. First I would like to ask Quentin Skinner, formerly Regius Professor of Modern History at Cambridge University, and recipient of the 2006 Balzan Prize for Political Thought, to speak to us about truth and explanation in history.

Quentin Skinner:

May I thank you all very much for allowing me to speak to you in my own language. This is a very great privilege, and Anglophones in particular must never take it for granted.

Truth and Explanation in History

Quentin Skinner:

When historians raise questions about truth, they generally do so in the context of talking about beliefs, and more specifically when talking about statements or affirmations of belief. I do not imagine, however, that historians necessarily have anything very deeply philosophical in mind when they talk about the truth or falsity of the beliefs they encounter in the past. When, as an historian, I find myself confronting

Balzan (ed.), *Truth in Science, the Humanities and Religion: Balzan Symposium 2008*,
DOI 10.1007/978-1-4020-9896-3_5, © Springer Science+Business Media B.V. 2010

some statement of belief, and when I say that I consider it to be true, I do not think that I am usually saying anything more than that I share the belief myself.

This is not to deny that a philosophically-minded historian might also have a theory of truth. Such an historian might have a view, that is, about what exactly they are committed to defending in saying of a statement of belief that it seems to them true or false. They may suppose that what makes a belief true is that it coheres with other beliefs. Or they may think that beliefs are rendered true by their correspondence to some facts about the world. Or they may not think that truth corresponds to anything at all. They may think that, when I say of a statement of belief that it seems to me true, and when I add that in saying this I only mean that I am disposed to affirm it myself, I may actually be enunciating a theory of truth.

I do not myself feel that, as an historian, I need to have any view about which of these theories of truth, if any, I should take to be the most illuminating one. However, as an historian of ideas I frequently find myself studying beliefs that people in the past affirmed to be true, and in examining these beliefs I cannot avoid noticing that, while some of them strike me as true, others strike me as obvious false and sometimes as completely incredible. Aristotle, for example, believed that bodies change quality whenever they change place; Thomas Aquinas believed that the sun circles the earth; John Locke believed that one can hope to observe stones growing in size. As any cultural historian knows, the list of such oddities is endless.

It is at this point, it seems to me, that one particular philosophical question about truth cannot fail to obtrude into the practice of the historian's craft. As historians, we are often interested in trying to explain why our ancestors held certain specific beliefs. The philosophical question we cannot avoid is then as follows: what role, if any, should be played in these explanations by our assessment of the beliefs we are studying as either true or false?

One answer that has lately been widely espoused by philosophers of history (and philosophers of social science more generally) has been that true beliefs need to be explained in a different way from false ones. Within the Anglophone philosophical literature, this view has been defended, for example, by Martin Hollis, Steven Lukes, Graham Macdonald, Philip Pettit, Charles Taylor and many others. The principal argument with which these philosophers have sought to uphold this commitment has been that the holding of a false belief points to some failure of reasoning, and that failures of reasoning require additional explanations of a kind that are not required in the case of true beliefs.

Consider, for examples, the discussion mounted by Macdonald and Pettit in their book *Semantics and Social Science*, in which they insist that judgements about truth and falsity cannot fail to enter into the explanations we give of the beliefs we encounter in past or alien societies. When a belief under investigation proves to be true, they maintain, no further explanation will be required. But when we encounter a belief which is 'manifestly false' or 'obviously incorrect', there is something further to be explained. We need, in particular, to consider what kinds of 'social function or psychological pressure' could have prevented the agent in question from recognising 'the mistaken nature of the belief'.[1]

[1] Macdonald – Pettit, *Semantics and Social Science*, 9, 34, 42.

It is worth prising apart – more fully than Macdonald and Pettit have done – the two distinct claims being made here. One states that true beliefs furnish their own explanation: when an historian encounters such a belief, nothing more than its truth need be invoked to explain it. The other claim is that false beliefs require to be explained in distinctively and strongly causal terms: what the historian is looking for in this instance, according to Macdonald and Pettit, is some kind of pressure or psychological blockage that prevented or inhibited someone from seeing and getting at the truth.

These assumptions have been widely embodied in the work of practising historians of recent times. By this I do not mean that the historians in question have necessarily been reading the philosophers I have cited. In fact that seems to me highly improbable. Rather the way of proceeding I have been outlining appears to strike many historians as the most natural one to follow.

To take a distinguished example, consider the explanation that Emmanuel Le Roy Ladurie gives in his classic study, *The Peasants of Languedoc*, for the upsurge during the era of the Reformation of beliefs about witchcraft. Ladurie begins by emphasising that the beliefs about witches held by the peasants were manifestly false, and were indeed little more than a product of what he calls "mass delirium".[2] If we wish to explain why these beliefs nevertheless gained such wide acceptance, he goes on, we need an account of what might have caused such a serious breakdown in normal processes of reasoning, and thereby caused the peasant consciousness, as he puts it, to break loose from its moorings.[3] The question, as he poses it, is what prevented the peasants from recognising the mistaken nature of their beliefs; what caused such an upsurge of obscurantism, such an epidemic of pathological beliefs.[4]

One element in the explanation offered by Ladurie is that, with the advent of the Reformation, the peasantry began to fear a loss of their traditional spiritual help. "Far from their priests, the peasants found themselves alone with their anxieties and their primordial fears – and abandoned themselves to Satan."[5] But Ladurie's principal hypothesis is that the peasants felt a deep sense of frustration at the collapse of the social upheavals associated with the Reformation itself. With the failure of social reform, their continuing desire to improve their lot took on a "mythical dress", and was forced to express itself in the "chimerical and fantastic revolt of the witches' Sabbath, an attempt at demonic forms of escape".[6]

I have now isolated two claims about the alleged importance of asking about the truth of beliefs in relation to explaining them, and I now want to comment on each in turn. The first states that, in the case of true beliefs, the historian need search no further for an explanation of why they are held. I have little to say about this argument,

---

[2] Ladurie, *Peasants of Languedoc*, 203–205.

[3] Ladurie, *Peasants of Languedoc*, 208.

[4] Ladurie, *Peasants of Languedoc*, 203–204, 206–207. Norman Cohn, *Europe's Inner Demons*, 258 makes similar claims about witchcraft beliefs as nothing more than a 'collective fantasy'.

[5] Ladurie, *Peasants of Languedoc*, 207.

[6] Ladurie, *Peasants of Languedoc*, 203.

however, for it strikes me as obviously untenable. To say that someone holds a true belief is not in the least to explain why they hold it, for they may have espoused it for reasons completely unconnected with its truth. They may, for example, have accepted it entirely on trust, as all of us are compelled to do with many of our empirical beliefs.

I mainly want to concentrate on the other claim I have isolated: that the espousal of false beliefs points to failures of reasoning that need to be explained in strongly causal terms. This is the principal argument put forward by the philosophers I have mentioned, as well as being the assumption governing Ladurie's explanation of witchcraft beliefs in his *Paysans de Languedoc*.

My own view – to state it at once – is that to follow this approach is nothing less than fatal to good historical practice. To do so is to assume that, whenever an historian encounters a belief which he or she judges to be false, the explanatory problem will always be that of accounting for a lapse of rationality. But this is to equate the holding of rational beliefs with the holding of beliefs that the historian judges to be true. And this is to exclude the possibility that, even in the case of beliefs that nowadays strike us as manifestly false, there may have been good grounds in an earlier historical period for holding them to be true.

It seems to me, in other words, that cultural historians need to operate with a strong distinction between truth and rationality. The reason is that, when we seek to explain beliefs that we judge to be irrational, it is at *that* juncture – and not at the point when we judge them to be false – that additional questions arise about how best to explain them. To equate the holding of false beliefs with lapses of rationality is therefore to foreclose – in advance of knowing whether this is appropriate – on one type of explanation at the expense of others. The causes of someone's following what are taken to be appropriate norms of reasoning will be of a different order from the causes of their violating them. It follows that, unless we begin by enquiring into the rationality of the beliefs involved, we cannot be sure of correctly identifying what needs to be explained, nor in consequence of directing our investigations along appropriate lines. If the belief proves to be one that it was rational for the agent to have held, we shall need to investigate the conditions of that achievement. If it was less than rational or palpably absurd to have held it, we shall need to enquire into the sorts of conditions that may have prevented the agent from following accepted canons of evidence and argument, or perhaps supplied the agent with a motive for defying them.

To illustrate the significance of these points, let me return to Ladurie's account of the beliefs about witches held by the peasants of Languedoc. He not only begins by noting that these beliefs were false, but his explanation presupposes that it could never have been rational to believe them to be other than false. Ladurie assumes, that is, that the falsity of these beliefs is enough in itself to show that they cannot have been rationally held. Operating on this assumption, he leaves himself no space to consider a quite different sort of historical explanation. He cannot allow that the peasants may have believed in the existence of witches as a result of holding a number of other beliefs from which that particular conclusion might reasonably have been inferred.

To consider only the simplest possibility, suppose that the peasants also held the belief – widely accepted as rational and indeed indubitable in sixteenth-century

Europe – that the Bible constitutes the directly inspired word of God. If this was one of their beliefs, and if it was rational for them to hold it, then it would have been the height of irrationality for them to have disbelieved in the existence of witches. For the Bible not only affirms that witches exist, but adds that witchcraft is an abomination and that witches must not be suffered to live.[7] To announce one's disbelief in the existence of witches would thus have been to announce a doubt about the credibility of God's word. What could have been more irrational than that?

Ladurie excludes in advance the possibility that those who believed in witches may have done so as a result of following some such recognisable chain of reasoning. But this not only means that he puts forward an explanation of witchcraft beliefs which, for all he knows, may be completely irrelevant. It also means that he bypasses a range of questions about the mental world of the peasants which it may be indispensable to answer if their beliefs and behaviour are to be satisfactorily understood.

A much more helpful way, it seems to me, of trying to enter into the alien world of early-modern beliefs about witchcraft is the one to be found in such classic studies as Stuart Clark's *Thinking with Demons* and, above all, Keith Thomas's *Religion and the Decline of Magic*. These historians proceed by treating beliefs holistically, asking how far it may have made sense to believe in witches by way of relating those beliefs to other elements in a much broader network of epistemic assumptions and commitments. What makes these studies so illuminating is their recognition of what this project involves. As they make clear, it will never be sufficient to show that the beliefs of those who accepted the reality of witchcraft were consistent with their other beliefs. It is also necessary to show – to put it rather formally – that they adopted their ideas about witchcraft in the light of a certain attitude towards the formation and criticism of their own beliefs. This is not to demand, on pain of irrationality, that they must have held certain specific beliefs. But it is certainly to demand that, if their beliefs about witchcraft are to qualify as rational, it will be necessary to show that they were concerned with the kinds of coherence, and where appropriate the kinds of evidence, that gave them grounds for concluding that their ideas about witches could be justified.

This kind of historical enquiry can lead to a variety of different outcomes. The historian may emerge with the conclusion that, although sixteenth-century beliefs about witchcraft were patently false, it was wholly rational to have held them to be true at the time. Another possible conclusion might be that it was only rational for such beliefs to be held with a certain and perhaps fairly low degree of probability. A rather different conclusion might be that, although the question of truth was not fully pressed, this was not in itself irrational – somewhat as Paul Veyne sought to argue in his book on whether the Greeks really believed in their myths.[8] Finally, the historian cannot I think exclude the possibility of concluding that the beliefs in question were not only false, but that there were no sufficient grounds for holding them to be true even at the time.

[7] See, respectively, Deuteronomy 13. 10–12; Galatians 5. 20; Exodus 22. 18.

[8] Veyne, *Did the Greeks Believe in their Myths*.

To insist on this last point is admittedly to tread on disputed ground. Stuart Clark, for example, argues that, once we uncover the inner acceptability of a given system of beliefs, we cannot fail to count it rational for the system to have been upheld.[9] I do not myself see, however, why our interpretative charity need be so boundless. But this is not in the least because I believe, in the manner of such philosophers as Martin Hollis and perhaps Donald Davidson, that there is a trans-cultural notion of rationality that can be applied as a kind of template to the past.[10] When historians stigmatise some particular belief as irrational, they need only be claiming that they have uncovered the prevailing norms for the acquisition and justification of beliefs in the community concerned, and that the belief in question was upheld in the face of, rather than in the light of, those norms themselves. They need not be claiming that the belief was irrational according to their own standards (still less *the* standards) of epistemic rationality. They need only be claiming that the agent in question fell short of – or perhaps abandoned, manipulated or in some other way deliberately defied – some accepted standard of epistemic rationality.

The essence of what I have been arguing is thus that, when cultural historians seek to explain the systems of thought prevailing in past societies, they should avoid even asking questions about the truth or falsity of the beliefs they investigate. The only point at which they should invoke the concept of truth is to ask whether our forbears had sufficient grounds for holding to be true what they believed to be the truth.

I am well aware, however, that anyone who argues in this way is bound sooner or later to be denounced (or commended) as a relativist, so I need to end by saying a word about whether I have indeed adopted a relativist stance. There is obviously a sense in which my argument is a relativist one. I have relativised the idea of 'holding true' a given belief. As I have suggested, it may well have been rational for the peasants of Languedoc to hold it true that there are witches in league with the devil, even if such beliefs no longer strike us as rationally acceptable. Furthermore, I have argued that all cultural historians need to be relativists in this limited sense. They need to keep before them at all times the thought that it is possible to hold a completely false belief with unimpeachable rationality.

Simon Blackburn argued in his keynote address that it is dehumanising to offer strongly causal explanations of people's true beliefs. The point I have been adding, speaking as an historian, is that it is no less dehumanising to offer similarly causal accounts of our forbears' false beliefs unless we are able to show in addition that those beliefs were irrationally held. Although their beliefs may not have been true, they may have had perfectly good grounds, by their lights, for holding them to be true.

---

[9] Clark, 'Inversion, Misrule and the Meaning of Witchcraft', *Past & Present* 87 (1980) 98–127, at p. 100.

[10] See Hollis, 'The Social Destruction of Realit y', 67–86; Davidson, 'On The Very Idea of a Conceptual Scheme', 83–98.

They may have imperfectly understood their world, but they may have been doing their very best to understand it.

It is a mistake, however, to suppose that cultural historians who espouse this position are embracing a thesis of conceptual relativism. Conceptual relativism is a thesis about the nature of truth: it contends that there is nothing more to truth that rational acceptability within a form of life. But this is not what I have argued. I have not asserted that it *was* true that at one time there were witches in league with the devil. I have merely affirmed that there may have been a time when it was rational to affirm it to be true that there were witches in league with the devil, even though such a belief nowadays strikes us as false. To put the point generally, I have merely observed that the question of what it may be rational for us to hold true will vary with the totality of our beliefs. I have never put forward the different and possibly self-defeating thesis that truth itself can vary in the same way.

I am not saying, in other words, that when Aquinas affirmed that the sun circles the earth, or when Locke affirmed that stones grow, these claims were (as relativists say) true for them although they are not true for us. I want to say that these claims were never true at any time. I am so far from being a conceptual relativist that I want to go much further and add that, as cultural historians, it does no harm, and may do a great deal of good, if we begin by focusing on such beliefs and the very fact of their falsity. False beliefs carry a special interest for cultural historians. For example, I find myself much more attracted to John Locke for knowing that he believed that stones grow. He must have been living, I begin to reflect, in a very different mental world, and it starts to seem a serious and worthwhile challenge to try to penetrate it.

Not only am I not a conceptual relativist, but I believe that the truth of conceptual relativism would be incompatible with the practice of cultural history. Unless there is some considerable overlap between what we believe to be true and what our forbears likewise believed to be true, and unless we additionally share with them some assumptions about how best to fit together our beliefs in order to construct arguments, then we shall have no means of gaining access to their world at all. The only point I have been making on the other side is that, in order to explain their world, we shall have to accept that they may have had good grounds for holding true a number of beliefs that we hold to be manifestly false. For example, that stones grow.

Panel Discussion

Nicolette Mout:

I would now like to ask Sir Keith Thomas to comment. Sir Keith Thomas is a Fellow of All Souls College, Oxford. He is a former President of the British Academy, and a member of the General Prize Committee of the International Balzan Foundation.

Keith Thomas:

Professor Skinner has asked how historians should proceed when they encounter persons in the past holding beliefs which they, the historians, regard as false. And he has provided an elegant and persuasive answer, with which I entirely agree. You might well think that, in view of his far too kind remarks about my own book, I could hardly do otherwise. In particular, I strongly endorse his view that it is just as necessary to explain why people in the past held what we regard as true beliefs as why they held false ones. In either case, the job of the historian is to ask what led people to think as they did. For example, Simon Blackburn told us yesterday that scientists think that the moon is about a quarter of a million miles away from the earth. You will be interested to know that I think that, too. But my opinion is totally second-hand. I hold it not because it's objective truth, but because I trust those who tell me that this is the case. I hold it because I take on trust what scientists tell me. The task of future historians of our society will be to explain why professional scientists enjoy such cultural authority, that most of what they say is unquestioningly accepted by the rest of us, incapable though we are of knowing whether it's true or false. Similarly, when historians study beliefs current in some periods of history, they should begin by asking who are the recognized custodians of knowledge, the accepted sources of authority at the time. By finding out who was entitled to tell people what to believe about matters of which they have no direct experience themselves, we shall get closer to finding out why they held the beliefs they did. This means that, if we are to understand why people in the past thought that some things where true and some were false, we need to study the history of child-rearing, education, religious teaching, literacy and reading, travel and all the other circumstances which helped to form the opinions they held.

I also agree with Quentin Skinner that people in the past often had good reasons for holding what we now regard as false beliefs. He cites the case of the philosopher John Locke, who once wrote that all stones, metals and minerals are vegetables – that is, they grow organically from seed. Now that may sound odd to us, but it was a view that had been widely held in the sixteenth and early seventeenth centuries, when many highly intelligent people believed that the universe was animate, and that physical matter was alive. This was by no means an irrational assumption. What, for example, could be more alive than fire, leaping and biting and gobbling things up? And some stones surely did grow – stalagmites and stalactites, gall stones, kidney stones. By Locke's day, this older belief in an animate world had been largely driven out by the new mechanical philosophy propagated by Descartes, Mersenne, Hobbes, Boyle and others. According to this view, matter was inert, and physical change the result of the motion of invisible particles, atoms or corpuscles. Fire, for example, was a mere tumult of corpuscles. Now, John Locke himself expounds this corpusculearean view in his essay *Concerning Human Understanding*, and it remains a puzzle as to why he should also have expressed the older doctrine about stones growing.

The explanation seems to be that he did so only once, in a little summary of the rudiments of science which he didn't write but dictated for use by a young gentleman whose tutor he was. This summary presumably dates from a relatively early phase in his career, before he'd fully absorbed the new mechanical philosophy. It was published some time after his death under the title *Elements of Philosophy* by an eighteenth-century editor who confessed that some readers had urged him to leave it out, for fear, as he put it, that the memory of Mr. Locke should suffer by it. But though the idea that stones grow does not represent Locke's mature view, it was by the standards of the time certainly not irrational, just rather old-fashioned. So the question which then remains to be asked is why had it become old-fashioned? Why did the mechanical philosophy triumph? And that is a question for the history of science.

Quentin Skinner is right, therefore, to urge us to distinguish between truth and rationality, and to point out that what seemed rational at one point in history did not necessarily seem rational in another. But that still does leave the historian with the difficult task of determining whether people were rational or irrational by the standards of the time if they happened to live in a time which didn't have a single accepted standard of rationality, but witnessed the coexistence of more than one set of what Quentin Skinner calls "prevailing norms for the acquisition and justification of beliefs". This was the case in John Locke's day, when the animate and mechanical views of the universe coexisted, but it is also the case in our own day. Are the pilgrims who travel to Lourdes in the hope of a miraculous cure behaving irrationally? Do they have sufficient grounds for believing in the possibility of such a cure? I suspect that not all this audience would agree about that.

The historian's task thus becomes the explanation of why different members of the same society think differently about what is or is not the criterion for truth. And that involves a great deal of research into their upbringing, education, experience, the experience of different social groups, people of different educational levels, and even the education experience and belief of individuals.

Nicolette Mout:

I now would like to ask Professor Dominique Schnapper, professor of sociology in Paris and recipient of the 2002 Balzan Prize for sociology, to comment.

Dominique Schnapper:

Let me just refer to the words of my friend and colleague, Skinner. He put me in quite an embarrassing position, because he has criticized a French historian through British historians, and I in some way have to reply to this criticism. I feel obliged to say I almost fully agree with his words. It could have been better for the sake of debate not to agree with him, but there is just one question I would like to ask him, a slightly critical one. You have given us a wonderful lecture about historical methodology. It would be extremely difficult to summarize your speech, but we could say it is the original sin of bad historians to try and judge the past starting from the present. My friend Skinner has given us a very good example of this. He has reproached Le Roy Ladurie for his interpretation about the past and its current beliefs. I agree with him. It's true that you understand much more about the past if you start asking questions about the reason why they were holding these beliefs, as there might be a good reason for having these beliefs. To understand them, we need to understand the reasons leading them to hold these beliefs. So we cannot just say that they were quite irrational. It's much better for a historian of culture to try and understand the past *Weltanschauung*, trying to distinguish not between what is true and false, but by trying to understand why they were holding these beliefs in the past.

We live in a democratic world, a world of immediacy. Historians very often say that history enables us to understand the present, which is not completely true. But trying to understand the present also enables us to try and understand another world. Using frustration as an explanation, as Ladurie did in his principal hypothesis on peasants and their belief in witchcraft, is a pseudo-psychological way of trying to explain it. Frustration explains everything and nothing at the same time. Let me just end with a question about your last remark: you historians of culture are not asking yourselves questions about truth. We might all agree on witchcraft, but do you think that a historian of culture can understand Galileo's problem without any intervention of the notion and concept of truth?

Quentin Skinner:

There is a moment when truth becomes extremely important for historians. We need to find out what matters most: trying to understand why somebody is holding a false belief – at least a belief which is false to me. This would be a very interesting approach. Historians are trying to find explanations, and this is the reason why the concept of truth is always extremely important and always present. But we were talking about explanation and the way to explain things. You need to have this concept of truth, but in my contribution I have stressed several times that it's really wrong to stretch this concept too much.

Dominique Schnapper:

There are several reasons leading me to thank the chairperson of this session because she has encouraged me to speak using my native language.

Cultural Relativism: Meaning and Limits

Dominique Schnapper:

The historians of the nineteenth century believed in the progress both of knowledge and of civilisation. Historians from that period held the Positivist view of the progress of knowledge by accumulation of 'historical facts'. Unlike them, in today's democracy – which, following Montesquieu, may be called "extreme democracy" – the prevailing idea is that historical science progresses less by an accumulation of facts and more by reformulating questions and interpretations. Does this mean we have entered the realm of relativism, and, if so, what kind of relativism?

It should be remembered that the "principle" of democracy – to use Montesquieu's words again – is "virtue", that is to say the upholding of laws and the subordination of the individual to society:

> What I call *virtue* in the republic is the love of one's country, that is to say the love of equality [...] I have therefore called *political virtue* the love of one's country and of equality.[11]

We should question the results of a trend towards a seemingly limitless equality which is never and may never be fully attained. It is in the nature of democracies that they should not be restricted by intrinsic limits. Like any political regime, however, a republic runs the risk of becoming "corrupt" – in Montesquieu's words – when people no longer act in accordance with its "principle". We may wonder whether the "providential democracy" in which we live is under threat, in accordance with its own internal dynamics, by "the spirit of extreme equality".

According to this spirit "of extreme equality", all opinions are equally legitimate. An opinion would have the same validity as knowledge, experience and thoughtful consideration. In a democracy "everything is an opinion" wrote Tocqueville. This may easily be seen when listening to the radio, where any 'opinion' – or 'first-hand account' in disc jockeys' parlance – is equally legitimate, given that all those expressing an opinion share in the same fundamental equality. Everything is equally valid. Relativism, therefore, expresses a shared ideology founded on the tendency towards indifferentiation between people and between orders: that of politics and of history, of justice and of equity, of analysis and of commitment… An individual living in extreme democracy will refuse to make value judgements, now considered derogatory. How can one speak of Truth, then? How can one still speak of Truth? Are research results in the social sciences just some opinions among other opinions or do they imply some form of truth? What particular kind of truth is the truth to which they aspire?

The Rise of Relativism

Relativism is a cardinal value of democratic society as well as being the bedrock of the social sciences. Anthropology in its widest sense stems from the desire to acknowledge and understand the diversity of human societies. However, the idea of a plurality of orders, of ideas and of systems is also part and parcel of the intellectual and political heritage of democracy. It is the distinction between the political and the religious order, between private and public life, between the legislative, executive and judicial powers which is specific to democracy and distinguishes it from theocratic or totalitarian regimes. Relativism is an asset shared by anthropologists and by democratic citizens. Just like any political regime risks becoming corrupt

[11] Montesquieu, *Esprit des lois*, avertissement de l'auteur, 227.

– Montesquieu's words again – so does relativism as a way of interpreting the world risk degenerating into ideology.

In order to understand today's rise of relativism it is therefore necessary to see it as one of the potential risk factors of extreme democracy. Modern democracy is based on the legitimacy of the community of its citizens, who are autonomous, free and equal. The dynamics of democracy tend to spread outside the strictly political sphere, into all areas of social life. The passionate aspiration to equality, in particular, that Tocqueville had already seen as the "begetter" of "democratic centuries", risks cancelling the distinctions between the various social roles and the different orders of collective life and, as a result, of unhinging all forms of collective organisation. Every society implies a hierarchy, entailing a form of labour division and a differentiation of roles, in order to organise a legitimate power regulated by shared laws, in order to manage those conflicts which inevitably arise in any collective form of life, conflicts which are a blessing in disguise since they represent the freedom of citizens. The "people" – in Montesquieu's words – is incapable of doing "everything unaided", "of deliberating on behalf of the senate, executing orders on behalf of the judiciary and of dispossessing all the judges"[12] without questioning the principle of a social organisation based on the rule of law. In our societies, which posit the equality of all as the foundation of their legitimacy, there is a constant risk that individuals, after legitimately demanding and gaining civil, juridical and political equality, will come to demand identity of all individuals. This demand would put into question any legitimate principle of differentiation and hierarchy thereby jeopardizing the very organization, legitimately regulated, of social life.

When democracy veers towards "becoming corrupt" – meaning that it no longer complies with the very principle of its organization – by becoming "extreme", autonomy may degenerate into refusing all forms of religious or political transcendence. In the past, this transcendence was embodied by the 'King' or by 'God'. In the early stages of modern politics they were replaced by another form of transcendence: the 'Fatherland' or the 'Nation' or the 'Republic'. In extreme democracy, direct interest and instant gratification tend gradually to become the primary goal in the existence of all individuals. Citizens' freedom risks 'degenerating' into license, that is to say into refusing to accept that freedom may only be exercised within the framework of rules acknowledged by all as legitimate. Equality would 'degenerate' into undifferentiated desires, into the aspiration not just to the civil, juridical and political equality of all citizens, but also to the identical and to the similar. As for relativism, *the* democratic virtue by definition, it also risks degenerating by not allowing any reference to the very idea of Truth and of Morals.

The three risks of "corruption" to which democracy is subject – the refusal of all transcendence, license and indifferentiation, all converge so that cultural as well as cognitive relativism may degenerate into ideology, in the form of absolute relativism.

---

[12] Montesquieu, *Esprit des lois*, Livre VIII, chapitre 2, 349.

Absolute or Relative Relativism?

Should the relativism to which we all subscribe, both as anthropologists and as democratic citizens, be relative or absolute? The very notion of relativism is something which has been metabolized by democratic societies. Modern developments in anthropology, in particular, have led ideas on the destiny of human societies to stress the relativity of cultures. This notion of relativity is not new. It was common even prior to the birth of the academic discipline known as "anthropology", from Montaigne or Pascal to Montesquieu, just to mention the French tradition. Scores of anthropologists have been inspired by the fight against ethnocentrism and the affirmation of cultural relativism, that is to say a doctrine that proclaims that all cultures enjoy the same validity. The 'great debate' raging among anthropologists and more widely among modern thinkers is not cultural relativism per se, which is commonly accepted not just by anthropologists but by society at large: it is more a question of establishing the meaning and the limits of such relativism. Should relativism be absolute or relative? Each culture is incommensurable, which makes it impossible to make any value judgment in so far as any value judgement is inextricably linked to the culture of those making such a judgement. Does this imply absolute relativism? Or rather, regardless of the relativity of cultures, is there a universal horizon authorizing moral judgements?

How should one answer the question put by Raymond Aron to Claude Lévi-Strauss in 1979: "Are universal judgements on moral behaviour incompatible with cultural relativism?"[13]

The debate of anthropologists on the universality of morals is not merely academic, lying as it does at the very philosophical foundation of policies adopted in respect both of native populations in countries created by European colonisation – North and South America, Australia – and of populations of foreign origin resident in European countries. In the view of absolute relativists, public policy should aim above all at preserving Native Americans in the United States or in Brazil from all contact since, as a result of the impermeability of cultures; any acculturation process would end up destroying them. In Europe's major countries of immigration like Germany or France, one should, therefore, allow such traditional features as female genital excision of girls from certain areas in Africa, or forced marriages, in the name of the equal validity of all cultures. Adopting instead the positions of relative relativists one should encourage a gradual and endogenous acculturation

[13] This question had been asked after a note by Claude Lévi-Strauss to the Académie des Sciences Morales et Politiques on October 15, 1979. The answer was as follows: "An ethnologist encounters both beliefs, customs, institutions which he may study, for which he can put forward a typology without any moral concern, and the people who created them and with whom he may have something in common. I shall not therefore try to answer the question. I should say it is an *aporia*, that we must live with it, try to overcome it by fieldwork, giving up, wisely, any attempt at a theoretical answer". Quoted in *Commentaire*, 15, (Octobre 1981), 372.

process of so-called native peoples in terms of the dominant modern civilisation, an acculturation which nonetheless does not negate cultural specifics as long as they are not incompatible with the freedom and the equality of all human beings. One should not refrain from passing moral judgements, condemning certain traditional practices which run counter to a universal idea of human nature, rejecting the idea that women be relegated to their biological dimension thereby losing the right to the full exercise of their freedom, which is what the practice of genital excision and forced marriages do.

Not all anthropologists accept the notion of absolute cultural relativism. The supporters of relative cultural relativism "consider the relativity of cultures as being itself relative, excluding neither the existence of absolute values nor the possibility of intercultural communication or the advantages of acculturation".[14] They base this choice on a theory of knowledge according to which the criteria of meaningfulness and of truth do not depend entirely on the cultural context, even though they are obviously not completely independent. Relative relativists renew the traditional criticism of scepticism since classical Antiquity: there exists a logical contradiction in the very idea of absolute relativism. In stating their theory, relativists imply that such a theory is true, *ergo* that there is a truth. By the very act of undertaking their research, anthropologists deem it possible to understand others – only to a certain extent, of course: after all do we know ourselves perfectly? Anthropologists cannot help admitting that, in spite of cultural differences, exchanges between people of different cultures are perfectly possible; they cannot help admitting that if not absolute at least relative knowledge *is* possible, or else what would be the point of anthropological research? An ethnologist, like all academics, considers that knowledge is preferable to ignorance, that the progress of science is, in and by itself, progress for humanity, meaning thereby that it is a universal value. How could one deem error preferable to truth? Is not tolerance towards others intrinsically superior to intolerance?

Against a parochial view of positivism – which must not be likened to Reason – it has rightly been observed that "historical objectivity has limits",[15] and more generally all knowledge reached through science, and through the social sciences in particular, has limits. However, this type of relativism is itself relative. It is true that historical facts do not exist outside the understanding given to it by researchers. It is true that researchers interpret the data available to them through their conceptual framework. Yet, the fact that all data imply interpretation should not be construed so as to mean that they are *simply* a matter of interpretation. There also exist true propositions. I shall keep to a few examples from the social sciences.

---

[14] Abou, *Retour au Parana*, 354.

[15] Aron, *Introduction à la philosophie de l'histoire.*

Partial and Provisional Truths

In so far as the social sciences go hand in hand with their practice they cannot be mentioned without also mentioning their production. Any consideration on the contributions of the social sciences and their relationship with truth must in fact imperatively base itself on research *practices*. To quote Weber, "methodology may never be anything but a reflection on the means which *arose* through practice, and gaining explicit awareness of this cannot be the foundation for productive work any more than the knowledge of anatomy can be for 'right' walking. Just like anyone constantly trying to control their way of walking on the basis of their knowledge of anatomy risks tripping up in the end, an academic may likewise be subject to the same mishap if she tried to determine the goals of her work on external foundations based on methodological considerations".[16] It is therefore by recalling a few results in the social sciences that I shall try to show the way in which truth and interpretation go hand in hand, which has characterised every stage of their development.

I would like to suggest that what they put forward is not Truth but some truths, which are themselves no doubt partial and provisional, in so far as they are subject to criticism by Reason and experience, whether it be experiments in the laboratory or historical experience of history. As such they can therefore be put into question by the discovery of new data and by advances in thought. They are, however, nevertheless, of a different nature from the opinions spontaneously expressed by the democratic individual who considers himself sole master of his own destiny and on which he bases his citizenship practices in order to justify them to himself and to others. Founded as they are on reason and on reflections on historical experience, these truths are renewed by surveys in the widest sense of the word. A survey may take several forms, not necessarily that of an opinion poll. The knowledge produced by historians, statistical data produced by public administration, observations and surveys led by academic in their specific fields are to be considered in this sense experiments. Like all enterprises in rational knowledge, truths in the social sciences are founded on experimentation.

It goes without saying that there is no history without historians, no sociology without sociologists, no jurisprudence without judges as we are often reminded by the so-called realist tradition of jurisprudence. It goes without saying that, philosophically speaking, there is no such thing as brute reality except in the sense that "everything is interpretation", as the saying goes. However, this should not be taken to mean that there is nothing but interpretation or that 'interpretation' means 'contingency' or 'improvisation'. Interpretation can only be given within the framework of knowledge of certain facts, it necessarily entails certain limits. I shall look at some examples from history and sociology.

[16] Weber, *Essais sur la théorie de la science*, 207–208.

History

Historical reason has been the subject of radical criticism since the end of the nineteenth century, in particular by Dilthey who showed how naïve historians were deluded into thinking that they could attain truth in the sense of reproducing the reality of the past exactly as it had happened (*wie es geschehen ist*), refusing to accept that the role of the historian consists in choosing from among facts depending on her questions, in producing a specific conceptual framework to understand and interpret them in order to give them meaning.[17]

Such criticism is today largely disseminated thanks to democratic values; the notion of relativism has gained the minds of historians in such a way that the very person who introduced historical relativism into the history of ideas in France, towards the end of his life suggested the idea that interpretation may not, of itself, provide a definitive answer to all questions: "Among Paris circles, the slogan "there are no facts" is still fashionable. Of course I am well aware that in a certain sense the slogan is philosophically true: all facts are constructs, the fact encountered by a physicist exists only within an intellectual system created by science itself. As for history, all facts are either perceived by the eye of the beholder or reconstructed by the historian from documents (…) but in the end (…) all societies are subject to *de facto* constraints, the need for production, for organisation, for a technological hierarchy, the need for a techno-bureaucracy and so forth".[18] After the ingenuousness of positivism in the nineteenth century, are we not seeing the excesses of relativism in the twentieth century?

Let us look at some examples and the progressive articulation of the partial and provisional truths which may be arrived at thanks to the work of the historian.

The Shoah *did* take place. The planned destruction of certain people – Jews, gypsies, homosexuals – was designed and put into effect in the name of an ideology proclaiming the superiority of certain 'races' and the damnation of other 'races, in particular the so-called "'Jewish race'. Gas chambers did exist, and they did gas people whose only fault was that they were born gypsies or Jews. Starting from this fact, which has nothing to do with interpretation, the development of knowledge poses endless questions, renewed by the historical experience of historians. What is the link between this monstrous project and the traditional Christian anti-Semitism of European countries? What was the effect of the humiliating defeat after the First World War on Germans and on their vote for Hitler? What was the role of Hitler's peculiar personality and what role did the system promoted by the Nazis play? By what means, manners and procedures did the final solution take place? How far were the different participants responsible, and how aware were they of what they were doing? These are only a few of the countless topics – among many others – which have given rise to research through debates by historians founded

---

[17]This criticism of positivism had been introduced and developed in France by Raymond Aron's thesis since 1938. Cf. R. Aron, *Introduction à la philosophie d'histoire.*

[18]Aron, *La révolution introuvable*, 122–123.

on rational arguments, from the critical scrutiny of sources, from reflections arising from the knowledge of other disciplines (like social psychology, for example), from a comparison with other totalitarian regimes. Facts are questioned in a different way when new documents or new reports are discovered, or when history leads us to raise different questions, for example when new genocides came to light or when the Soviet archives were discovered after the implosion of the Empire.

These later interpretations may certainly make the events more intelligible and yet, it goes without saying, we may not give indisputable or definitive answers to these questions, and to others which may arise in the future. There still remain, for example, all kind of other questions on the singularity of the Shoah: is this singularity absolute or relative? If by all evidence it is something unique, like any event, it may, it must be compared to other genocides of the twentieth century. This comparison may have seemed shocking to a number of Jews for whom the desire to destroy the 'Jewish people' by the Nazis seemed a sacred fact. Accepting this view would be a misunderstanding of the social sciences and their goal of rational knowledge. A comparison does not imply identity, nor even proximity. It is, on the contrary, a rational method to try and understand similarities and dissimilarities. For example, the bureaucratic nature of the Shoah seems to be specific to it. It raises the metaphysical question of the application of technical and scientific reason, the glory of Western civilisation, to the execution of a monstrous plan. Is the cold-bloodedness of the plan what distinguishes the Shoah from the outbreaks of violence of which history has given us so many examples? And were not the Khmer Rouge deportations of an equally systematic nature? Or is the number of people killed what characterises the Shoah, the plan to make an entire people disappear from the face of the earth? Or rather the dehumanisation plan which went hand in hand with the executions? Or rather a unique combination of these different characteristics? Here too, by making a comparison, interpretations may be enriched and yet it is not possible to put forward some absolute truth. However, some partial and provisional truths may still be established which are destined to be 'revised' in the future.[19]

Such truths do not exhaust reality, nonetheless, and this for three significant reasons. On one hand it is inevitable that historians, being themselves historical individuals, should end up asking the questions that present reality suggests to them. Historians in the future, starting out from a different experience, will ask different questions of the past and will analyse the Shoah in a different way.

On the other hand, the series of events that historians know, while by definition that very past was unknown to its protagonists, has given a different meaning to the events of the past, both of the collective past and of the past of single individuals. The essential nature of François Furet's work has consisted in a rereading of the French Revolution in the light of the revolutions of the twentieth century and in particular of the Soviet revolution of 1917. Stalin's deportations suggested a new interpretation of the Terror during the French Revolution. Was it just a question of

[19]From among a wide array of works the following may be noteworthy as an example: Sémelin, *Purifier et détruire*.

mishaps to be considered the price to pay for the Coming of New Man and of – at last – a just society? Or rather was the Terror of necessity part of a radical plan to rebuild a society from scratch and does this as a result question the legitimacy of the idea of a radical revolution set on building a society from a blank slate?

Lastly and more importantly, historical events, be they individual or collective, are, after a fashion, complex, ambiguous and inexhaustible: the interpretation of such facts knows no end. Equally in this sense, there is no "end of history", interpretations will always be subject to revision. Historians are by definition 'revisionists'. And yet the legitimate plurality of interpretations, renewed by the experience and the development of views is not at all the same thing as revisionism.

Sociology

Turning to a lighter subject, one of the abovementioned partial truths that can be put forward is that a liberal economy is more effective at producing goods and services than a government-regulated economy. This truth was brought about by considering the fiasco of communist economies and the economic success obtained by a country which continues to neglect human rights but which has chosen a liberal economy because it is more effective, namely China. This truth does not, however, in any way undermine the philosophical consideration that may be made on what it means to accumulate goods and services, a questioning of what it means for people to act as a society. GNP is neither a gauge nor an indicator of how happy a people are, but this enables us to ask questions as to the meaning of modern societies starting from a few true propositions. Since it is a question of understanding society, one cannot follow the "masters of suspicion" (Raymond Boudon's definition), from Foucault to Derrida, who have insisted on the irrationality of mankind and affirmed the radical relativity of all knowledge. Truth in the social sciences is partial, or possibly partial, as suggested by Weber with his "ideal type", but it nonetheless *does* exist.

Let us take another example which concentrated sociologists' minds in the period from 1960 to 1980: inequalities in educational results as a function of social origins. In France data produced in a rather anodyne way after surveys by the National Institute of Demographic Studies (INED), under the leadership of Alain Girard, at the beginning of the 1960s had not led to much debate. It was on the interpretation of these true facts that a large number of sociological works were later conducted. Some sociologists interpreted these data as a sign of privileged social categories trying to keep their dominance over popular classes in order to secure to their own advantage the 'reproduction' of capitalist society from which they benefited. This was represented, albeit with some major differences, in the analysis by Pierre Bourdieu and Jean-Claude Passeron,[20] and later by Christian Baudelot and Roger

[20] Bourdieu – Passeron, *Les héritiers; La reproduction.*

Establet.[21] Raymond Boudon, for his part, later demonstrated that these educational inequalities were fewer than those in the social sphere, and that as a consequence education, unlike what the previous authors had suggested, instead of consolidating and legitimising social inequalities, tended rather to diminish them, and that, in any case, it played a rather limited role.[22] Historians, on the other hand, showed the often perverse effects of measures intended to fight social inequalities at the educational level, as was the case in particular of the creation of comprehensive secondary schools.[23] The research underscoring such arguments stirred passionate debates among sociologists in the period 1960–1980 and has recently been taken up again.[24] The school system has profoundly changed over the course of the past decades, educational inequalities have appeared under new guises and the social system itself as a whole has evolved, sociological knowledge has made progress: this means that new perspectives are required. Starting from the indisputable fact that educational inequalities exist, interpretations change not merely depending on the philosophical conception that different sociologists have of society but also on the historical evolution of education and society. No analysis is final.

One could perhaps suggest that the school system contributes at one and the same time to the 'reproduction' and to the 'mobility' of democratic societies and that one should try to appreciate, if not to gauge, the role played by the one and the other, the synergy of the one and the other. It should be kept in mind that in a society as a whole, equality is a principle, an ideal, a regulatory idea and not a reality; society sets itself tasks which are impossible to fulfil completely. It should also not be forgotten that on such highly charged topics sociologists adopt positions which are always, in part at least, ideological. All these reservations are part of the research itself – all research must be self-critical – but they must not obscure, on one hand, the brute reality that educational inequalities as a function of social origins do exist, and, on the other, that sociologists from one generation to the other have produced analyses of different aspects of the educational system and of the role that education plays in democratic societies, which constitutes itself undeniable progress in the advance of rational knowledge.

I would like to add a few words about the so-called realist school of jurisprudence, which, in the same spirit of absolute realism, affirms that judges are free to give the texts that they interpret whatever meaning they may choose. Indubitably, from a rigorous philosophical standpoint, there is no law except in so far as a judge applies it. Philosophically speaking there is no text unless there is someone interpreting it. It is true that the interpretation of a point of law consists in discovering the meaning of a text, which means that any decision entails a creative aspect.

---

[21] Baudelot – Establet, *L'école capitaliste.*

[22] Boudon, *L'inégalité des chances.*

[23] Prost, *L'enseignement s'est-il démocratisé?*

[24] See the works by Dubet, in particular *A l'école*, as well as those of Van Zanten, in particular *L'école de la périphérie.*

No point of law is ever crystal clear and absolutely applied, a number of interpretations are possible. However, the interpreting faculty also entails some constraints. To suggest that interpretation is all there is to it or that interpretation is synonymous with improvisation or with contingency would mean yielding to a peculiar form of absolute relativism. The text itself may require interpretation to such an extent that the interpretation becomes weak or even inexistent. The interpreting faculty is exercised within a framework of legal constraints defined by the texts and by jurisprudence. Although it may be true that "several interpretations of a point of law are possible, from among which a choice is made (the creative faculty of the interpreter, namely the judge), there are nevertheless certain interpretations that are excluded by the very drafting of the point of law, which are simply impossible".[25] Judges too make pronouncements by formulating partial and provisional truths, truths which, however, cannot disregard the law and jurisprudence.

The fact is that sociologists, like historians, hardly ever manage to establish objective analyses and that they have a tendency to confuse the pursuit of knowledge with the expression of their passions and even with a political struggle. We must never forget the inevitable limits of sociological objectivity as well as of historical objectivity, due without a doubt to the weaknesses of researchers and, more radically, to the fundamental ambiguity of social and historical phenomena. But whatever the weaknesses of researchers – and why just researchers in the social sciences? – we must extend the history of historical understanding and of sociological thought and their aspiration to rational knowledge, we must continue to exercise critical reasoning, that is to say self-critical analysis.

The democratic world, subject as it is to immediate needs, is constantly threatened by the rule of peremptory statements, of short-termism and of murderous slogans. Against this heavy trend it behoves those who wish to devote themselves to "in-depth, slow work" in Tocqueville's words, to fight in favour of patient, honest and often effective research, and to make the voice of this research and of rational knowledge heard. The partial and provisional truths in the social sciences contribute in their way if not to Truth at least to the aspiration to and the search for truth.

Panel and General Discussion

Keith Thomas:

Professor Schnapper has given us a marvellously robust and wide-ranging talk, and I shall make only a few brief remarks on what she says about history and historians. She accepts that all historical interpretations are constructs, patterns imposed upon past events, which will vary according to the historian's point of view. And she rightly reminds us that no historical interpretation is final, because new perspectives

[25] Hamon, *Les juges et la loi*, 220.

will inevitably put the past in a different light. But, as she also points out, that doesn't mean that anything goes. All interpretations are not equally valid, and it is not just their rhetorical power or our own ideological preferences that make us regard one interpretation as superior to another. What we seek from interpretations of the past is plausibility. They must be consistent with the surviving evidence and with our own assumptions about how human beings behave. Of course, that still leaves much room for disagreement, since the surviving evidence will almost certainly be consistent with more than one interpretation, and our assumptions about how human beings behave will reflect the presuppositions of our particular culture. Moreover, the sad fact is some historians, or so-called historians, are more honest than others, less inclined to suppress inconvenient facts or to misrepresent the evidence. So all historical interpretations must be approached critically. But that is the nature of life.

In our everyday existence, we have to pick our way through good witnesses and bad ones, and we recognize that we can never establish more than a reasonable probability that things were the way we think they were. But though the interpretation and re-interpretation of past events is a never-ending process, we do know beyond any reasonable doubt that many of these events did occur. Many historians spend little time on interpretation and devote their energies to establishing and discovering new facts about the past, whether they are historical demographers calculating the size of past populations, or economic historians plotting the international movement of capital, or architectural historians trying to discover who designed which building and when. They all agree that, though the past will never yield all its secrets, a good deal more can be found out about it. Of course, these so-called facts about the past are only facts because our conceptual schemes make them so, hence Nietzsche's assertion that there are no facts, only interpretations. We can't know everything about the past. We can't begin to know the whole truth about the past.

For example, it would be impossible to give an adequate account of what has happened in this room during the last few hours. It's possible, of course, to record everything that's been said, and no doubt that has been happening. But how much of what has been said has each of us heard and understood? And what about the miscellaneous thoughts which have been going on in each of our heads, some of them wholly unrelated, no doubt, to the official proceedings. Most of what has happened this morning has gone forever, without leaving any trace behind. Experience is infinitely fuller and richer than any historical account of it can ever hope to be. So if we cannot hope to recapture all the dimensions of an event at which we were present, how can we hope to know much about what happened centuries ago, when it's left no trace at all?

Nevertheless, it is in the belief that we can find out more about the past that most historians pursue their vocation, and they do so in the confident knowledge that, as Professor Schnapper rightly stresses, many historical claims about the past can be objectively established and leave no room for argument. At the peace conference at Versailles in 1919, a German historian said to the French Prime Minister, Clemenceau, "I wonder what later historians will make of all this", to which Clemenceau replied, "Well, at least they won't say that Belgium invaded Germany".

Nicolette Mout:

I would like to ask Professor Mauro Baranzini, professor of economics at the Università della Svizzera italiana, the University here in Lugano, to present his comments.

Mauro Baranzini:

Professor Quadrio Curzio should have exposed his ideas and points of view, and he asked me to take his place at very short notice. I'll try to do my best to explain what happens and what happened in economics, which is my discipline. I think the issue of truth in economics has got to be set against the object and methods which are used in economics – that is to say, to consider the tools of analysis which economists use to explain economic facts. We have had three important schools. The first one was the classical school. In 1823, David Ricardo, said that the laws which regulate the distribution of the products between socio-economic classes made up the most important aspect of economics. And Ricardo maintained that, provided we've got labour enough to produce, the production of commodities would not be scarce, but abundant. Then in the 1870s came the neo-classical school. In 1933 Lionel Robbins, summarized the goal of the school and said that economics studies human behaviour as a relationship between ends and scarce means. So scarcity returned as being one important factor of the object of analysis in economics. The neo-classical school also asserted that full employment would always be ensured in the system, provided the state did not intervene and provided other factors were present in the system. Then came the economic crash of 1929, the Depression of the early 1930s in America and in Europe. And in Cambridge, John Maynard Keynes wrote his general theory of employment in 1936. For him, economics is a science which has got to study the mechanisms which must ensure full employment of labour and capital. It is a normative approach to economics, while the neo-classical school did not have a normative but a positive approach to economics.

It is clear that the truth in 1929 and 1930 was that there was a high rate of unemployment, and that the previous economic school was not able to explain this fact. Now, about 70 years after John Maynard Keynes has written his book on full employment, there is a new school, a new generation which has chosen a much more pluralistic approach characterized by cross-fertilization in economics. The three schools are brought together, the three tools of analysis are combined in order to see whether it's possible to explain better what happens in the real world. Problems like economic growth, sustainable growth, income and wealth distribution, the fight against poverty, better distribution of wealth and income have been tackled and new solutions have come forward. In sum, I have to say that economics is a relatively new science, or discipline, and therefore we are just at the beginning of our task.

The role of truth in social sciences and economic sciences in general is not easy to identify or to describe, for different reasons. First of all because economic science is a relatively new science, and secondly because economists have still not agreed on the method and object of political economy. As I've said, three great schools of economic thought have so far emerged for scholars to work with. The first, the classical school headed by David Ricardo who, in 1823, wrote that the laws regulating the distribution of income among social classes were the most important object of economics. Afterwards, in the 1870s, as a reaction against the thesis of the classical economists, the neo-classical and marginalist school came to the fore, which advances that, especially at an economic level, scarcity dominates economic systems and that economic science is concerned with the study of relationships between scarce goods, commodities, and limited needs. Among other factors, this school has always advanced that, in a particular situation, when the destabilizing intervention by the state is absent, full employment should always be ensured automatically. The 1929 crash and the early 1930s Depression demonstrated that at least this was not true. Keynes, who put forward the third most important school of economic thought, wrote his book in 1936. His fundamental assertion is that economic science must devote itself to the study of those mechanisms which may guarantee full employment to economic systems through external intervention of an actor, in some cases a state, in other cases someone else.

I stressed that a new generation operating after the 1960s and 1970s has tried to combine the three different schools of economic thought, trying to cross-fertilize them by facing the problem using a plurality of ideas and approaches, and it seems that – I say this in all modesty – the results have been quite good. An economic science or discipline with a cross-fertilization of ideas may help us to understand better the social and economic phenomena of our time.

Nicolette Mout:

We now have time for questions from the audience, and the questions could be for any one of our speakers.

Question:

I'd like to put a question to Professor Skinner about his very impressive summary of the problems of interpretation in history. It seems to me that all the stories that you tell are correct, except that something has been more or less left out, which is a sort of starting point: how do we pin down the beliefs that the people are supposed to have? You said that we shouldn't go for truth, but rather for rationality and the grounds and evidence that people have for their beliefs, but you didn't ask the question how do we interpret, or how do we in the first place localize something as a belief that they had? In general, there are three ways of knowing that. The first one is to know the causes of the belief, the second one is to know what people said, and third maybe would be to go to their brains, which are unfortunately inaccessible. It seems to me very difficult to pin down and to interpret the beliefs of, say, some person in the past, whether it be a peasant from Languedoc or a famous philosopher, without noticing at some point that the belief has a certain content. Now if the belief has a certain content, the question is whether this content is true or not. Of course there is no external point of view on truth, as Bernard Williams said that there would be no sort of external standpoint on reality which would make it possible to see whether a belief is true or not. But of course we have to judge the truth of the belief starting from our own lives. We have to compare our own beliefs about what's true with what we take other people think to be true. It seems to me that we cannot do that without using the notion of truth itself. I want to point to a sort of basic problem which has been elaborated by many philosophers. I would say that Le Roy Ladurie was not completely wrong, even though he fell into the trap of attributing the wrong reasons for believing in witchcraft to the peasants in Languedoc.

Quentin Skinner:

Considering the three possibilities you offer me, there is of course built in to what I'm talking about a kind of deep assumption that people affirm what they take to be true. That's a kind of Davidsonian thought about getting interpretation going. Of course, we don't think that's always the case, not just because of hidden codes like irony, but for many possible reasons we might want to say that in general that's the case. However, as historians we've got no option but to go with your possibility 2. Notice that I'm talking about affirmations. Questions about truth don't usually come up in relation to beliefs, they come up in relation to statements of beliefs, and I'm talking about the appraisal of those statements. The reason I want to avoid questions about truth – and I'm really sorry not to have persuaded you of the Anglophone view of Le Roy Ladurie – is that I'm talking about what it is to be a rational credent. That's all.

You might want to say that what I've maintained about what it is to be a rational credent (insofar as I sketchily laid it out) is connected with questions about consistency and evidence. If you want to say: that's what it is to track the truth, then that's fine by me, but that's just a *façon de parler*.

Dominique Schnapper:

Probably he would not agree!

Quentin Skinner:

I've said nothing that's false.

Question:

But you didn't tell the whole truth.

Dominique Schnapper:

At the end of your paper the implication was that you just don't consider the problem of truth. And that's not true! I mean, you don't tell the story in the same way when you talk about witchcraft or science...

Quentin Skinner:

What I consider is a story that some people talk about, but I'm saying that it's not a story about truth; it's a story about what it is to have rational beliefs. Now of course you want to say that rational beliefs track the truth, and maybe that's a way of talking. But I can give an account of what I think it would be to hold a belief in a rational fashion that's not going to talk about the truth. The reason that I don't want to talk about the truth is that to ask questions about the truth of these affirmations that interest me in the past is to move in a particular explanatory direction, and that I want to avoid. I want to hold that much more open.

Dominique Schnapper:

Yes, but even keeping it open, at a certain moment, you ask yourself if it's true or not.

Quentin Skinner:

No, I don't.

Dominique Schnapper:

Oh yes, you do! You don't talk about science as you talk about witchcraft.

Quentin Skinner:

I do, I do!

Dominique Schnapper:

Well, you're wrong then.

Nicolette Mout:

Well, I did hope that the sisters wouldn't quarrel, but now they are quarrelling.

Quentin Skinner:

Just whom are you applauding?

Nicolette Mout:

Both, as I have to be impartial today.

Question:

I would like to contribute an illustration to the story of *Les paysans de Languedoc*. You have given a wonderful explanation how it was totally irrational for the people to believe in witchcraft, but it does not explain why it happened there, and why it happened then. Why not in the Rhone valley, or during the Hundred Years' War? It was localized, it was limited in time, it had all the ingredients of an epidemic. There's a hypothesis that there was an epidemic, a blight of the rye, a sickness of that staple which makes it poisonous. One of the poisons involved is lysergic acid, a precursor of LSD, which is hallucinogenic. Now if most of the women start acting even funnier than they usually do, it is quite conceivable that, given your historical explanation, the idea of witchcraft comes up. Why only women, or preferentially women?, It was totally rational at that time to feed the women with bread while the men ate the meat. I don't think it can be proven to be true, if there is a truth. But it is certainly an example showing how the combination of science and humanities can lead to deeper insights.

Question:

I would like to ask a question that I have already asked in a number of seminars in which the Anglophone world has been better represented than the Francophone one. I'd like to speak of the concept of the imaginary. In Skinner's analysis, he essentially took into consideration two categories: the rational and the irrational. The concept of the imaginary occupies not just in France but in other European countries as well an important place in research concerning the very function of the imaginary in relationship with memory, rationality, individual memory, collective memory, and so on. Professor Changeux explained the functioning of the brain this morning and he gave representations their true place in the brain. The English language does not have this concept: the imaginary. In this symposium which concerns itself with truth, I would like to have an explanation of this lack of the concept, or this indifference to it.

I would also like to make some remarks about a concept which until now has not been mentioned. The word 'Islam' has so far not been heard. I'm not talking about the millions of Muslims outside Europe, but about those in Europe. What about those Muslims who claim to have the opportunity to apply Sharia law to co-believers living in England or France? What does it mean to claim that Sharia law can be applied, a law which stems from a certain belief? Is it a false belief that the Sharia is of divine origin, whereas it is not, as lawyers and historians have demonstrated? But how should we qualify this? How are we to write the history or the sociology of belief? How do we qualify this in terms of people, Muslims, who are citizens in our society? It is in danger of becoming a political problem, political management of what is true and what is false. This is not just an intellectual responsibility, and if politicians depend on what answers intellectuals or historians and political scientists come up with, the problematical aspect of truth is laid bare, also in our daily life in society, which has already been exposed to several tragedies.

Nicolette Mout:

We'll go into this during the next session.

Session V: Theology – Religion

17 May 2008 – 2:00 p.m.

Chair:
Sergio Romano

Speakers:
Georges Cardinal Cottier OP *Faith and Truth*
Mohammed Arkoun *Violence, the Sacred, and the Regimes of Truth*
Geza Vermes *The Truth about the Historical Jesus*

Panel:
Jan Assmann; Brian Hebblethwaite

Sergio Romano:

The theme of this afternoon's session is theology and religion. We have been lucky to be able to call around this table people who, because of their personal and academic interests, are somewhat representative of great monotheistic confessions. The Catholic Church, represented by a cardinal; the Islamic world, represented by an islamologist. Reformed Christianity is represented by a scholar who is a minister of the Anglican Church, and we also have an Egyptologist who will perhaps talk to us about the relationship between Egyptian religion and monotheistic religions. The first speaker is Cardinal Cottier, who is from Geneva. He entered the Dominican order in 1945, was ordained in 1951 and became a professor in Geneva. In 1989, he was elected secretary of the International Theological Committee, and in 1990 he became Pontifical Theologian. As Titular Archbishop of Tullia he was created cardinal by Pope John Paul II in 2003.

Balzan (ed.), *Truth in Science, the Humanities and Religion: Balzan Symposium 2008*,
DOI 10.1007/978-1-4020-9896-3_6, © Springer Science+Business Media B.V. 2010

Faith and Truth

Georges Cardinal Cottier OP:

I

Introduction

Christian conscience has an essential relationship with truth. Christian conscience is located in the extension of Israel's alliance, which is an alliance with the only God, namely the only true God, who has made Himself known to His people. Compared to Him, the gods of the other nations are but vain things, idols made by human hands: they are false gods. An intrinsic bond, or rather an identity, can be seen to exist between God and truth. God is truth.

It is God who, of His own free initiative, has made Himself known to us, who enlightens us about His mystery. The acceptance in us and by us of this light is faith.

This, in a few words, is the theme that I would like to develop.

The *Prologue* of the Gospel according to John

The *Prologue* of the Gospel according to John ends with the following: "No one has ever seen God; it is the only Son, who is close to the Father's heart, who has made Him known." (John 1:18).

No one has ever seen God. Sight is direct knowledge, immediate, face to face. Such a sight is not within the power of any human being, who is infinitely distant from God. What is hereby stated is the Transcendence of God. God is, for a limited spirit, inaccessible, incomprehensible, unnameable, not because He is without a Name, but because this name surpasses all other names. One cannot therefore get hold of this name, which calls for the religious silence of adoration. The affirmation of divine transcendence lies at the foundation of the whole process. Starting from this affirmation one may put the question, which shall be examined later on: can our reason know something about God?

Nevertheless, we do have access to the Inaccessible; we can, without in any way diminishing the first affirmation, pronounce the ineffable Name. This is what we do when we speak of the "only Son, who is close to the Father's heart".

In the *Prologue*, it is essentially a question of this "only Son"; He it is who leads the way, who is Himself the way (cf. John 14:6). He is presented from a dual point of view: that of divine eternity and that of the history of salvation.

Let us start from this second aspect: "for the Law was given through Moses, grace and truth have come through Jesus Christ" (John 1:17). This phrase marks a breakthrough. Moses, before receiving the law from God, had received the revelation of His name: "I am he who is." [others translate: "I am that I am"] (…) "This is what you are to say to the Israelites, "I am" has sent me to you." (Ex 3:14). Notice that, in the Gospel according to John, Jesus takes on himself this name: "I am", "Ego eimi".

Grace and truth mean achievement and fullness which are given through the Incarnation. Earlier on we had read: "The Word became flesh, he lived among us, and we saw his glory, the glory that he has from the Father as only Son of the Father, full of grace and truth". (John 1:14).

The Word – such is the name that the Gospel according to John gives to the only Son – is at the heart of history; the past is headed towards its coming; the present and the future live in the expectation of its final glorification, which shall also mark the end of history.

The Word

The Word, then, is the name of the only Son. Let us now consider the first aspect. In fact, from the start the *Prologue* has been telling us what is the identity of the Son.

> In the beginning was the Word: the Word was with God and the Word was God (John 1:1).

The "beginning" (*archè*) establishes a parallel with the opening of *Genesis*. It refers to the origin, the eternal birth of the Son, before all time and before history.

The Latin *Verbum* is a translation of the Greek *logos* which itself corresponds to the Hebrew term meaning "word". Both the First and the New Testament use the term "word" to tell us that God speaks to humankind. This divine word is effective and works in different ways. It is a creative word, so that the creature sings the glory of its Creator. Its interventions, due to the sovereign initiatives of the love He has for its People, mark the great moments of a history guided by its Providence. It speaks to us also through the prophets and ultimately through its Son (cf. Heb 1:1–2).

Let us note here that the meeting between the Word, such as it is understood in the Judaeo-Christian tradition, and Greek thought, in which the *logos* plays a central role, is at the heart of Western culture and, as we shall see, of any culture encountering Christianity. For its part, in fact, Greek thought will give the *logos* wide semantic developments. Referring at first to a pronounced word, an utterance, *logos* will then take on the meaning of inner word and later of reason, thought. The fecundity of the theme of *logos* in Greek philosophy is well-known. However, before dealing with this point, let us return to the *Prologue*.

Revelation

"The only Son (...) who has made Him known." (John 1:18). We have at first focused on the statement: "The only Son" who is the "Word". By naming him thus, it is true, the Gospel gives a foretaste of the nature of his action: He "who has made Him known". This may equally well be translated as: "has told, has unveiled". In other words: He has revealed Him, He has made Him manifest. The Son brings the revelation to its fullness.

"No one has ever seen God": not because a possible meeting did not take place, but because no-one, that is to say, no created intelligence can see Him, is capable of seeing Him. Paul will make this explicit: he announces that "What no eye has seen and no ear has heard, what the mind of man cannot visualise"… this is what God has revealed. It must be borne in mind that Paul is here speaking of the "wisdom of God" which he contrasts with the "wisdom of the world" (cf. I Cor 1:17–2:16). Greek philosophy, for its part, is a quest for wisdom. The contrast highlighted here is one between two approaches to wisdom.

Revelation is not the result of human effort, it is not self-made revelation. It is a light received, a gift accepted; it takes its origin in a free initiative of God. Revelation makes manifest of God that which is inaccessible, it makes it known. That which is revealed and which, according to Christian faith, will be fully revealed after death, has therefore a special status in our mind. Our mind, which receives it and gives its adherence to it is not, however, on an equal footing with it. That which is revealed and to which the believer has access does not lose its transcendence – which would otherwise be the greatest distortion and the greatest imposture, but remains enveloped in mystery – a mystery which enlightens and pacifies, a matrix for reason. This is the locus, both of Christian mysticism and of Christian theology.

The mystery denotes the status which the object of revelation has in the mind of the believer. Nevertheless, such a status is temporary and will yield its place to the final one. The shroud enveloping the mystery will be torn apart and we shall be face to face with God. The First Epistle of John says: "My dear friends, we are already God's children, but what we shall be in the future has not yet been revealed. We are well aware that when he appears we shall be like him, because we shall see him as he really is." (1 John 3:2).

Testimony

The only Son tells us of the mystery of the Father. He may do so because, being Himself close to the Father's heart, He sees that which we cannot see by ourselves. Our access to mystery rests on the mediation of the incarnate Word.

Jesus Christ, who is Himself the object of the revelation, is the one who reveals. He is called by a term that makes this meaning explicit: "the faithful witness" (Rev 1:5).

In the transmission of revelation bearing witness plays a fundamental role.

Our intellect aspires to seeing its object unmediated; it aspires to evidence. There, it finds its perfection. In most cases the unmediated nature of evidence is not possible. Mediation is necessary. Reasoning reaches an apodictic conclusion when it shows the necessary tie linking this conclusion to principles immediately apparent in their evidence. Not every demonstration has this perfection, it is nevertheless possible to establish its degree of probability and the quality of a plausible hypothesis.

As a way of access to truth, testimony has a special nature. In fact, here someone else has seen on my behalf that which was not possible for me to see, someone else knows and intends to share with me that which he knows, not by demonstrating it for me, but by personally making himself guarantor of that which he is announcing. He takes a personal stand, since he asks me to put my trust in him, to rely on his truthfulness, to believe in what he says.

A reasoning is false whenever an error creeps in; a testimony is false whenever the witness deceives me.

Before accepting such testimony it is thus appropriate to ask ourselves some honest questions. Is a witness credible, in his person, in his actions, in his behaviour? In other words, a witness must present his credentials. For my part, I have to ask: must I or can I believe the testimony that has been proffered? Credulousness is the precipitous fall for the acceptance of something which offends reason.

Having examined what is known of the witness and of his actions, an enquiry comes down to the following question: are the pieces of information and the intelligence collected sufficient in the eyes of reason? Is it reasonable to believe?

Credibility

The foregoing holds for any testimony, generally. But in the case of a revelation axed on divine realities, a new and essential consideration is essential.

Since it is a question of a purely human testimony, I can, by way of thought, put myself in the place of the witness: that which he has seen I could myself have seen, or another person could.

In the case of the revelation of mystery, which is by itself inaccessible to me, the only Son alone can reveal it to me, since He is close to the Father's heart. This means that He Himself, in that which constitutes his identity, is shrouded in mystery. Once again, this means not only that it would be reasonable, as in the case of testimony on purely human things, to put trust in Him, since He has a certain weight and plausibility and because He could by rights be verified by other means. Here testimony bears on "hidden things", supernatural things, so that the question becomes: is it reasonable to believe in supra-reasonable mysteries?

The issue of credibility has an existential value, since it concerns the meaning of destiny as far as its dimension of eternity. It is an issue that encompasses one totally. After inspecting its credentials, the mind can but better gauge the transcendence of mystery. Saint Paul speaks of foolishness (cf. 1 Cor 1:17 f.) – the foolishness of the initiatives of divine love. Pascal has admirably underlined this paradox: in the eyes of the nature of divine grace which impels us to believe, our religion is foolishness; in the eyes of the signs by means of which it shows its divine origin, it is wisdom.[1] Proof is human, faith is a gift of God.

It would be appropriate here to consider the distinction, introduced by a number of exegetes, between the Jesus of history and the Jesus of faith, a distinction which

---

[1] Cf. Pascal, *Pensées et opuscules*, 587.

some have pushed as far as rupture and opposition. The meaning and the limit of such a distinction are discussed in the beautiful book by Joseph Ratzinger Pope Benedict XVI, *Jesus of Nazareth.*[2]

Let us content ourselves with noticing that this is an epistemological problem. The type of question that is asked of history or of the text is crucial in terms of the answer given. This has nothing to do with subjectivism, rather with an awareness that the question depends on a point of view and is not therefore exhaustive, and it is also a question of considering what Rudolf Bultmann has called pre-understanding.

Theological Thought

Credibility means the credentials presented by the message revealed in order to be judged acceptable by reason. Faith, which is a gift of God, opens up the understanding of the mystery. This is why the New Testament, when speaking of testimony, mentions the decisive role played by the testimony of the Holy Spirit, the foundation of certainty and enlightening. Faith makes knowledge possible, it has entered the mystery. This is what Jesus says to Philip: "Anyone who has seen me has seen the Father, so how can you say, "Show us the Father"? Do you not believe that I am in the Father and the Father is in me?" (John 14:9–10).

The link between seeing and believing is arresting. Faith, for all its obscurity, is knowledge. Far from demanding that reason be left by the wayside, faith is for knowledge a source of a new-found vitality, as expressed in Anselm's well-known formula: *fides quaerens intellectum*. The mystery does not radically waylay reason. It is super-understanding, an excess of light with regard to its natural capacities.

Reason, reflecting upon itself, knows that it is not itself its own foundation but that it participates of a higher reason, which is reason in its fullest meaning – divine reason. It discovers at the same time its own limits and also an urge, a longing, which impels it to rejoin its origin. By virtue of this original kinship, and of the aspiration which is congenital to it, once it receives the revelation of the divine *logos* it finds itself stimulated; it is a new awakening. Hence Anselm's second formula, complementing the first one: *intellectus quaerens fidem.*

Theology thus corresponds to a natural requirement of the believing intellect. It strives to penetrate, by means of the conceptual tools at its disposal, the intelligibility of the mystery, to discover its hidden harmonies, to illuminate with its superior light, and to unify its knowledge of the world, of humankind and its history.

Thus reason operates a two-way intervention relative to faith. Its intervention is required at the preparation stage: it is the gauging of credibility; the life of reason is directly stimulated by faith itself, attracted by the light of mystery: such is the *intellectus fidei*, i.e. theological thought.

[2] Ratzinger, *Jesus of Nazareth*.

Truth

The believer gives his consent to God who leads into His mystery by means of revelation, he gives his consent to Him who is the first truth.

Faith is not a subjective feeling; it is acceptance of the first and founding truth.

Truth is understood first and foremost as the correspondence between reality and the idea that I have of it, since knowing is the becoming of the thing intentionally.

Knowledge is reached when, reality being perceived by understanding, I know at the same time the truth of that which I am affirming of it, namely the adherence of my judgement with the reality as it is known. Such is the meaning of the definition of truth, as adherence of the thing and of understanding: *adaequatio rei et intellectus*.

There exists therefore a constitutive kinship between understanding and reality. If the destiny of understanding and the aim of its activity is to know reality, this presupposes that reality responds to its expectation. Being is knowable.

Correspondence means conformity and that to which one conforms is essentially a rule. Knowledge is true when it corresponds to reality, in which it finds its measure. Our understanding is true when it is measured by things apprehended in different degrees of amplitude and depth. Capturing just one aspect is true only in so far as this measuring conformity takes place.

If things are capable of exercising this measuring function to such an extent that the understanding that knows becomes intentionally the thing known, this is because, as we have seen, being is knowable.

The Light of Being

Here, we are here touching on the second meaning of truth, which is in any case presupposed by the first. Truth designates now the being itself in so far as it confronts understanding as the measure to which it conforms.

This is why things, since they do not have their being from themselves, receive intelligibility, which is at one with their being, from the source of all intelligibility.

They receive from the divine intellect which conceives and creates them, their intelligible configuration. As their being is shared, so is their intelligibility. Therefore, if things measure our understanding, this is because they themselves, in so far as they are, are measured by the One who, being the first intellect, is the origin of all intelligibility.

In God there is identity of being and intellect. He Himself is truth and the origin of all truth. By receiving from Him their being, the other beings participate, each in its own degree, of His truth, they are themselves intelligible, capable of being known by created intellects. Our intellect reaches the perfection of its act once it realises its correspondence with those known beings that are its measure. But these beings are a measure only by virtue of being themselves measured by the creative intellect.

Faith

Through revelation God communicates directly to reasonable creatures the truth that He Himself is. In order to speak Himself, through the prophets and through the Son, he uses our human language, but he does not go through the mediation of created things that allow us to know Him as their First Cause.

Faith is acceptance of the word of God, consent given to direct testimony of the first and original Truth introducing us into his mystery.

In so far as he is the Son of God, Jesus Christ, who by virtue of his humanness is the way, can say of Himself: "I am (...) Truth and Life". (John 14:6). And further on we read: "And eternal life is this: to know you, the only true God, and Jesus Christ whom you have sent" (John 17:3).

II

"The greatest of the Christian truths", wrote Pascal, "is the love for truth". This is an inspiring and awesome statement.

It is awesome when one thinks of the political use to which it may be put and to which it has in fact been put, as well as of the forms of intolerance and fanaticism for which it has been used as a pretext. The fact that such aberrations have often been used as an argument for a relativist position begs the question of its anthropological meaning. Can man be understood without a constitutive relation with the Absolute? The highest aspirations as well as the craziest adventures have their explanation in such a relation. Conversely, what is the properly human meaning of an existence enclosed in what is relative? Whenever we call on human dignity we are affirming the transcendence of the person in relation to the surrounding universe. This transcendence in its turn has its roots in the transcendent Absolute.

Revelation and Philosophical Thought

The encounter of Christianity and Greco-Roman culture was a major event having exemplary consequences. As a question of determining the ultimate end of existence, the opposition between Christian wisdom and pagan wisdom affirmed by Saint Paul is not called into doubt. However, at the same time, any compromise with idolatry being excluded, the predominant attitude is one of sympathy, being ready to accept any speck of truth, seeing it as a half-way stage on the way to encountering the fullness of truth. Christian theology, for its part, will make use of the resources of a highly advanced philosophical thought in order to move on with its own developments.

Such an approach does not depend on contingences, rather it rests on the conviction that grace does not so much destroy nature as lead it to its flourishing. It is in fact the same God who is at the origin of natural reason, which proceeds by means of its own ways and methods, and at the origin of revelation. By rights there cannot be opposition or conflict between natural truth and revealed truth.

This principle seems to have been belied by history more than once. In every case, so to speak, a lack of respect for a epistemological rule is observed; it is exactly the rule of respecting the sphere appropriate to each area of knowledge, while being aware of its specific nature and of the limits of its field of application.

Revealed truth sheds the supreme and the most comprehensible light. It is the principle of a unity of sapiential order. This means harmony, balance, integration, a respect for differences and articulations; it cannot replace those forms of knowledge which it unifies by acknowledging their specificity.

The theologian referring directly to revelation fails in his mission if he intervenes precipitously or if he insists on imposing conclusions while ignoring forms of mediation.

A long-lasting effort of research and exchange of opinions may well be necessary before agreement and synthesis are reached.

This is why, fundamentally, it seems to me that the problem can only find an answer at the epistemological level.

Truth and Violence

Fanaticism brings us to the question of the relationship between truth and violence. Does recourse to violence constitute a legitimate means of defending truth?

When Pope John Paul II invited Christians to ask forgiveness for errors which, in the course of the history of the Church, have constituted, instead of the expected testimony, a counter-testimony and scandals, he was guided by a conviction which is shared by the Church and which concerns its historical dimension: time is given to us not just so that we may personally proceed towards saintliness, but also so that the style and customs of the Christian community as such may be ever more in conformity with the requirements of the Gospel. This is only possible because we are made capable of conversion and of the "purification of memory".[3]

This said, an ever more vigorously affirmed principle clarifies the question: the freedom of the act of faith. It is only freely that I give my consent to the first Truth, after loyally examining the credentials and under the spiritual movement of grace which never forces conscience. Personal engagement is part of the act of faith,

---

[3]See my *Mémoire et repentance.*

so that an adherence extracted by means of force or obtained as a result of threats, of psychological pressures or of corruption is not an adherence of faith.

A second principle intervenes, the principle of the evangelical distinction between the things of God and the things of Caesar (cf. Mt 22:21). This principle does not mean that the business of the secular city is no longer subject to regulation by objective moral law, as if to open the door to political amorality. It witnesses to the nature of evangelical law as the law of the Kingdom of God, which is distinct from the secular city. In other words, human destiny does not find its ultimate achievement in time, it is trans-historical it finds its achievement in eternity. A Christian is thus a member of two cities. The requirements of the city of God, although they have some effects, in terms of inspiration, on the management of human affairs, are transcendent in relation to those of the political realm. They are condensed in that charter that is the *Sermon on the Mount*. The better this essential distinction of the two orders is perceived, a distinction which is neither opposition nor separation, the more awareness one will gain of the fact that the affairs concerning the Christian community as such must be conducted in accordance with a way and a style which cannot be identified with methods having their legitimacy in terms of secular societies.

The Defence of Truth

This said, truth must be defended against all forms of lies and deception and, equally, against errors. It is legitimate that those who lack the resources of knowledge and culture be protected against specious discourse. Modern states have made us aware of the extent of ideological domination. The different forms of monopoly in advertising and propaganda which are characteristic of the media show that democratic societies themselves are not totally protected from pressures exerted on people's minds.

It can be seen that this question pertains essentially to the morality of the media and their conformity with the essence of the Christian message.

Vatican Council II has declared the liberating principle which must guide conduct in the defence of truth. Speaking of conscience, the *Dignitatis humanae. Declaration on religious freedom* (1965) states: "The truth cannot impose itself except by virtue of its own truth, as it makes its entrance into the mind at once quietly and with power."[4]

Trust in the force of truth is an eminent sign of faith. This force has found its most eloquent expression in the testimony of martyrs and of all those who have stood firm in the face of totalitarian persecutions. It is them, more than their executioners, that history remembers.

---

[4]http://www.vatican.va/archive/hist_councils/ii_vatican_council/documents/vat-ii_decl_19651207_dignitatis-humanae_en.html, n.1.

Sergio Romano:

Cardinal Cottier gave us a quotation to the effect that truth does not impose itself by violence.

Violence is precisely the theme of Professor Mohammed Arkoun: violence, sacredness and truth. He was born in Algeria and has taught in different universities in France, in the United States, at the Pontifical Institute of Arab Studies in Rome, and was a Fellow at the Wissenschaftskolleg in Berlin. He is currently research director and member of the board at the Institute of Ismaili Studies.

Violence, the Sacred, and the Regimes of Truth

Mohammed Arkoun:

> However, the number one obstacle to the search for light is quite probably the will to power, the desire to show off one's virtuoso abilities or to provide a shelter against too evident objections. Truth is a limit, a standard which is higher than individuals, most of whom harbour a secret animosity against its power.
>
> André Lalande, *Vocabulaire technique et critique de la philosophie*, 'Préface' (Paris 1926)

I Introductory Remarks

The theme chosen for this symposium requires that we should go back to several traditions of belief, thought and knowledge. I am aiming at all forms of religious, cultural and scientific truth, not just at the doctrinal guardians of religious Truth supporting the orthodox faith, or their rivals who have since the Middle Ages opposed the postulates of classical metaphysical Truth. I am aiming also at those secularised citizens, men and women, who are regularly called on to vote and give to the democratic state the required legitimacy to exercise the monopoly of legal violence according to Max Weber's definition. This legitimacy covers all spheres of life: political, administrative, economic, cultural, educational and so on. Truth, called *alétheia* in Greek, meaning that which is not hidden and referring to the recurrent idea of 'unveiling', is present in all forms of religious thought, and reappears – with other tools for research and reflection – in modern thought. Here, God is one object of knowledge among others, a hypothesis which is no longer necessary to the learned, nor to politicians, nor even to free thinkers.

As a historian of Islamic thought, I have come across the issue of religious Truth with a capital 'T' in a confrontation with the scientific and philosophical truth in the classical phase of the history of Islamic thought (750–1300). Paradoxically, the fertile confrontations with thought from this distant age have come to an end with the triumph of repetitive traditionalist thought that remained indifferent if not hostile to the rise of modernity in Europe until the present day. This paradox has become a collective tragedy in the second half of the twentieth and at the beginning of the twenty-first century through the fundamentalist forms of religious 'truth'. Another paradox also emerged in Europe with Europe-wide wars between countries which have contributed considerably to the enrichment of scientific revolutions, intellectual criticism and artistic creativity. Nazism, like Stalinist Communism, has indicated forms and contents of Truth which refer back to radical questions concerning the functioning of the human mind and, above all, reason in the most civilised and advanced contexts.

My ideas as a researcher and teacher have considerably benefited from rich, powerful, passionate and always instructive debates which have marked intellectual, scientific, cultural, artistic and political life in France during those exceptional years from 1960 to 1980. Innovative works have been published in several fields of knowledge and culture. Eminently instructive epistemological tensions have kept researchers and teachers spellbound, so much were the latter supposed to convey to their public new trends in meaning, new cognitive strategies. These subverted the thought and value systems that remained active, providing more fertile alternatives until the fall of the Berlin Wall. At the beginning of the 1990s, voices from the United States announced 'the end of history', pointing in fact to the era of a distinct polarisation of power, economy, free market expression of all forms, levels and modes of truth.

My obsession while writing my doctoral thesis in the 1960s was that of applying to Islamic studies[1] all the new methods and problematisation techniques introduced in history, sociology, anthropology, psychology, psychoanalysis, literary criticism, linguistics, semiotics, as well as semiology. This was before 'the end of history' and the triumphant theory of the 'clash of civilisations' lost their previous dynamic cross-disciplinary practice and returned to their fragmented specialist nature. History, in particular, disputed the field to linguistics and anthropology. The *Annales* school, installed in the new *Ecole des Hautes Etudes en Sciences sociales* under the direction of Fernand Braudel, pursued with the help of talented and fertile minds the enlargement of the historian's field of study. New explorations of as yet ignored territories became familiar under the aegis of the appropriate disciplines such as historical psychology, sociology, anthropology and even epistemology. These cognitive strategies helped to identify the important concept of 'truth regimes' for the first time. Jacques le Goff, Georges Duby and their disciples assigned to Medieval history new tasks defined as 'the anthropology of the past and the archaeology of everyday life'. This programmatic expression was illustrated by countless works of quality not just about the European Middle Ages, but about all periods of history to our present time. Students took part in this revolution of mentalities and of thinking tools as they drew on promising intellectual and scientific responses and proceeded to reappraise all inherited certainties, values and knowledge, as evoked by the demonstrations of May 1968.

As a participant in this ebullient period I was too weighed down with developments in research and teaching in the wide field of societies and thought that were far behind the times. Such developments were closely linked to the cognitive, existential

[1]The term then used in France was 'islamology', a term now made obsolete by the widespread academic use of the term 'political sciences' following the model and the geopolitical mapping of Good and Evil under the indisputable *Machtpolitik* of the United States.

and ideological treatment of what I have termed the 'Qur'ânic fact' and the 'Islamic fact'. I declined straightaway the umbrella term 'Islam' invoked by so many people of all origins and of all adherences, a term which has become dangerous even when used by researchers, essayists and teachers. In 2008 the manipulations of this term are even worse, as they carry with them all forms of Islamic radicalism and of those reactions which are supposed to eradicate it. Even at the Sorbonne I had to be careful with my teachers who kept themselves aloof from what they called 'Parisian trends', stigmatising the ephemeral nature of these ideological rumblings. I have, however, kept my freedom to choose each year the themes and the spirit of my lectures. I have inaugurated two fields of research and teaching to make room not so much for detailed knowledge accumulated in works on Christian, later secular, European societies, as on the methodological and epistemological debates separating different schools in each discipline. It was necessary to identify data which were specific to my research field: Islamic thought. This term, which I had to defend, was not self-evident, since each field of study was divided into specific disciplines such as theology, law, exegesis, philosophy and literature. In 1973 I outlined for the first time the tasks and the cognitive strategy of what I called 'applied islamology' as distinct, on one hand, from 'classical islamology' based on the philological study of the great ancient texts, and, on the other, from the political sciences that were then rapidly gaining ground.[2]

In my books, with provocative titles, I collected a number of articles published between 1970 and 2003 on two strategic fields in the history of Islamic thought: (1) the 'Qur'ânic fact', and (2) the 'Islamic fact' considered at the different levels of the use of reason, always in confrontation with the primacy of the revealed data and of the contents of so-called orthodox faith. *Lectures du Coran* focuses on the linguistic and semiotic plurality of methods and approaches to deciphering Qur'ânic discourse before imposing any theological or exegetical statement on it. It is a question of liberating the reading of the Qur'an as a text from all the presuppositions inherited from the traditional interpreting communities. The aim is to bypass the heresiographic hierarchy recognising only one orthodox community and rejecting all others as sects condemned to eternal punishment. I thereby opened in the field of Islamic study the domain, at the time seldom visited, of an 'archaeology of knowledge' and of a 'sociology of failure or of success' of works, authors and schools of thought. As it happens, the social frames of knowledge, which are subjected to regressive beliefs and conducts simultaneously advocated both by the post-colonial single-party states and by the Islamist opposition movements to which they have given rise, have rejected and continue to place obstacles in the way of all overtures put forward in my book *Pour une Critique de la Raison Islamique*. Even among the circles of researchers and teachers there are partisans of the islamisation of modernity, opposing everything that comes from what they call the 'cultural aggression of the West (Westoxication)'. The sociology of failure becomes

[2]In 1982, I published my *Lectures du Coran*, in 1984 *Pour une Critique de la Raison islamique* and in 2002 *The Unthought in Contemporary Islamic Thought*.

paramount at all times and in all contexts where the triumph of the guardians of the orthodoxy of Islamic faith and Law is manifest.[3]

Each self-proclaimed orthodox community or politico-religious movement brandishes the flag of the authentically received, interpreted and lived Word of God, thus implying the damnation of all deviants or deniers. An example by a reputed historian, infuriating because of its dogmatic excesses, has recently been given in a book in Arabic by the Tunisian Mohammed Talbi.[4] An analysis of the articulation of Truth in this book, helps us gauge the impact of uncritical religious belief on the system of the individual and collective subject in all contemporary so-called Islamic contexts. Everything confirms the intellectual, cultural and scientific relevance of the cognitive intervention strategy suggested in my publications, though it is very little known and understood, both in what I call the 'societies of the Book' and in those societies now dominated by the culture of unbelief.

This is why I have continued to work for the enrichment and adaptation of my research programmes both in my writings and teaching as well as in public conferences throughout the world. I have tested the reception skills of my various audiences from Indonesia to Morocco, from Moscow to South Africa and from the whole of Europe to North America. I have looked for the most diverse and linguistically and culturally the most distant audiences in order to augment the concrete data without which applied islamology cannot exist. I have been pursuing this methodical search ever since 1965 to this very day. My participation in the present Symposium in Lugano is part of an observation and an experimentation which enrich and consolidate my cognitive strategy of intervention in research proper (a dialogue with my colleagues) and in sharing, through oral exchanges, the reliable results of critical knowledge.

II Religious Truth, Philosophical Truth, Scientific Truth and Common-Sense Truths

My doctoral thesis, devoted to Arab humanism from the fourth to the tenth century,[5] has led me to question the various aspects of reason in Islamic contexts. This is why I returned to the issue of humanism in a second book 36 years later.[6] In the programme and articles published earlier, I had concentrated especially on the so-called classical period of the thought and civilisation qualified as Islamic. The uses of reason in contemporary Islamic contexts (1945–2008) refer to areas

[3]Many such examples will be found in my latest book with its explicit title: *Pour sortir des clôtures dogmatiques.*

[4]*li-yatma'inna qalbî* [For the peace of my heart].

[5]First edition 1970; 3rd edition, 2005.

[6]*Humanisme et islam.*

little frequented by researchers, but which are nevertheless very fertile ground for the examination of issues regarding the reversibility of cultures and of civilisations, the regression of the status and of the products of reason in a phase of the general history of thought and of the sciences, which, in Europe at least, continues to favour the development of critical reason as a concept both innovative and liberating for the human condition.

This quick review of an intellectual itinerary clearly shows that the issue of truth underlies all 'Combats et propositions' evoked in the subtitle of a second book. This book was devoted to revisiting the humanist issue in its totality in contexts in which a militant fundamentalist version of Islam, today's or yesteryear's, is predominant. "All writing is an act of historical solidarity", was the motto of Roland Barthes and of those linguists attached to the practice of what is called 'discourse analysis'. Every definition, other critics say, is a takeover from reality and the appropriation of discourse to the aims of the will to power, often under the guise of the will to know the truth and to live according to its imperatives. Whoever says 'power' moves away from authority, nearing the constraints imposed by reason as categorical imperatives of intelligence, of morals, of the exercise of the monopoly of legal violence. Aware as I am of the slippery slope of all discourse towards power and violence, I use the term 'propositions' proffered for continuous debates within the community of researchers/thinkers, not just of the partisans of cold and factual erudition. If critical discourse articulated by reason is offered as a series of propositions to be debated with partners believing in this same self-control of reason, then it may be possible to work together to overcome the involution, the oblivion, the omissions and the drifting by the wayside to which any speculative activity is subject. Such considerations have given rise to my programme 'of a reflective history of Islamic thought' whose methodologies, developments, conceptualisations and rhetorical apparatus look for validation not only within the specific sphere of Islamic thought but also within closer cognate spheres (Jewish and Christian thought with their several ramifications in the history of monotheism, itself inseparable from its Mediterranean anthropological basis), and more distant ones (Chinese, Hindu, Buddhist and African thought and their respective anthropological bases).

III The Concept of Emerging Reason

The French historian Paul Veyne has recently published a book devoted to his colleague and close friend Michel Foucault.[7] One may be surprised that he should have waited so long before sizing up the contribution and the personality of such a great spirit departed only too soon. One will be even more surprised at the subtitle which refers to the banality of the history of ideas, of which the methodological

---

[7]Veyne, *Foucault*.

shortcomings have rightly been stigmatised by the very same person who struggled most to get rid of the illusions of 'historical-transcendental themes' in order to let reflexive thought take root in concrete life situations of social actors of all times. When asked about this aspect, Veyne revealed that his subtitle had been 'The Samurai and the goldfish': two allusions too obscure for a title. Still, the goldfish refers to an episode in Foucault's life when he was seven. As his mother was about to present him with a goldfish in a bowl, he asked the following, philosophically-charged question: "Mammy, what does a goldfish think?" Later Foucault would show that people turn around like goldfish in a bowl filled with pseudo-truths. He completely altered the approaches to truth in the thought systems inherited from the past. The concept of emerging reason became imperative in my mind during a long confrontation between the historical vicissitudes of reason in Islamic contexts after the installation of the Imperial Caliph State (661–1258) until the emergence of post-colonial states in the 1950s on the one hand, and the conquering, critical and reason-creating dynamism in Christian and later modern contexts in Europe on the other. The comparative study of the two destinies of the dialectic of the *logos* and of the *mythos* on the one hand, and of the Word of God and religious discourse on the other, enriches and radicalises considerably the critique of the works of reason, of the conditions of its successes and of the forces which make it regress until it is submerged by collective fantasies and the alienation of critical faculties.

In this constantly enlarged perspective, to speak of 'post-modernity', as it has been done in a fortunately fleeting fashion, means getting locked up in the purely European/Western track of reason. Emerging reason instead, as it is endlessly on the look-out in all cultures for novel challenges in human history, widens and multiplies its spheres of experience, the sources of its information and the rebuttals that come from all sides. In my talk I shall enlarge upon the new and liberating tasks of emerging reason by making use of the lessons of the ideological polarisation 'Islam versus the West'. I shall examine this pair, which arose out of polemic and liberation wars, in the long, medium and short term. It will be seen that the issue of truth will move towards more radical questions, more restrictive critical imperatives as well as more humanistic forms of communication.

In all cultures, the quest for truth implies reason coming to terms with all the questions, all the methodologies, all the validation operations linked with the production of 'sense', which must always be distinguished from 'that which makes sense' in such and such discourse situations and not in other situations. 'Sense' can only be produced if one bewares of facile slips towards dangerous 'effects of sense'. In linguistic and semiotic analysis reference is also made to 'the sense-destroying genesis' and therefore to the destruction of the truth(s) conveyed by sense which may be nothing but an 'effect of sense' and therefore of the truth(s) which crumble(s) when put to the test by the discourse analyst. This is why substantial, intangible and eternal truths, constructed in the course of the Middle Ages through the postulates of the revealed Word or through the tools of Aristotelian logocentrism, prove to be fragile and unsustainable once they have been subjected to the deconstructions of all thought systems. Such 'adventurous truths' continue nonetheless to support the religious beliefs and metaphysical speculations of idealist thought. It is true that such truths continue to have the function of refuge for all those who have no access to research and analytical tools used by the smaller research community. These functions of religious Truth are clearly illustrated by the resistance of Orthodox Christianity to the official atheism of the Soviet regime, by the angry reaction of so-called Muslim peoples against colonial regimes and the *Machtpolitik* of the Big Four, in particular in the Middle East after 1945. Judaism as well as Islam have continued to be used as a refuge, springboard and source of legitimization for the political actions and recurrent wars of the past 60 years. Buddhism and Hinduism are also not exempt from such functional degenerations of religious Truth which finds itself ever more involved in combat ideologies. Their expansion has come to corrupt great democracies like the United States since the attacks of 11/9/2001 and the reactions these have produced.

The same can be said for scientific truths validated by clinical trials, laboratory tests, mathematical calculations and all other forms and ways of experimenting. It is well known that in this area, too, there have been some spectacular distortions that have exposed even human dignity itself to the worst dangers. The few powers possessing the most advanced technology insist on the legitimacy of what they call 'clean' wars, that is to say without human losses, unlike the traditional Europe-wide wars. Weapons and the conditions in which they are used do not seem to be within the range of ethics committees in the life sciences. Such political practices, which are common since the frequent resort to terrorism, have devastating consequences for the responsibilities of science and of scientific reason cut off from ethics as well as from economic and financial reason.

The cognitive sciences have dealt an even more decisive blow to what we continue to call the truth: the good, the ugly, the beautiful. Jean-Pierre Changeux has recently produced a bulky volume comprising a synthesis of his research and theories since his *L'homme neuronal* (1983).[8] He attempts to open up new pathways to the search for truth by moving from the molecule to the soul. We are confronted with data and fundamentals which are both more reassuring and more worrying, since the speculations and the abstract and fragile conceptual constructions are abandoned in order to establish the functions of language, of knowledge, of intelligibility, of judgement and of value on data arising from clinical observation. Under such conditions it may be worthwhile to consider the concept of 'intellectual adventures' in the sense of vagaries or adventurous coherences of reason. Maurice Merleau-Ponty had earlier examined 'the adventures of dialectics' both in philosophical terms (Hegel and Marx) and in terms of the ideological distortions of Leninism-Stalinism. Here, too, the writing of history has its dangerous excesses, comparable in all respects to those of the theologians interpreting so-called Holy Scriptures. We have seen the havoc wreaked by the positivist historicism of philosophies of history in the service of the will to power (cf. the recurrent debates on the 'clash of civilisations' and 'the end of history' immediately after the crumbling of the Soviet and Maoist enemy). As fundamental a discipline as philology has nurtured the arbitrariness of

---

[8]Changeux, *Du vrai, du beau, du bien*.

academic mandarins: I remember the 'imperialist' tendencies of linguistics in the1960s at the Sorbonne under the pretext of leaving behind the crimes of philologism. These brief reminders show that history has always contributed to the construction of what each *epistéme* holds as true on the basis of representations, beliefs, definitions, collective images which place harmful limits on spirits, bodies, everyday relations, individual and collective conducts. Thus, 'written history' holds stakes in truth that do not always adequately encompass the truths of 'history as it is lived'. The debates on the methodologies and epistemologies of each writing of history are concerned with this distance between, on the one hand, the reductions, the selections, the cutting up, the categorisations and the arbitrary interpretations of the historian and, on the other hand, the truths of life as it is experienced by each social actor and each collective subject.

After publishing my *Critique de la Raison Islamique* in 1984, I have constantly tried to make explicit the order of reasons and the concrete practical objectives – in particular in terms of the transmission of knowledge – on which this *Critique* is founded. It extends the field of historical investigation by introducing the sociology of the 'thinkable' and of the 'unthinkable', of the 'thought-of and of the unthought-of in the logospheres' which coexist in the same political and social space. The higher the number of controls exercised on the thinkable in a logosphere considered at any given historical moment, the wider the space and the number of unthinkables. The same is true of the pair thought-of/unthought-of. Thus the constructions, functions, circulation of 'truths' become complex and bear witness to the precariousness of each proclaimed truth. This terminology, which is ordinarily restricted to philosophical thought, has reached out in my writings to key disciplines needing deconstructing, reflexive, archaeological and cross-disciplinary analysis of all oral and written discourse, of all the corpora and anthologies, of all cultural codes, and of all semiological systems. By spurring the *Critique de la Raison Islamique* along such ambitious and complex paths within past and present Islamic contexts, I wanted and still want to inscribe 'the Islamic archive' – in the sense developed by Jacques Derrida – in the comparative history and anthropology of all systems of thought claimed by religious, national and community traditions.

In spite of my repeated explanations, a large number of my colleagues still introduce me as a Professor of Islamic thought, or, which is worse, as a Professor of Islam! At the Sorbonne, I found it difficult to persuade the academic authorities to name my chair 'history of Islamic thought', so ingrained was the current practice of favouring time-honoured designations such as the special fields of philosophy, theology, literature, law and so on. It is well known that in the Middle Ages such frontiers were not watertight, especially in that period that saw the emergence and expansion of what I called the 'philosophical *adab*'.[9] One of the most important challenges of truth in my struggles to illustrate and introduce a new outlook in Islamic studies, concerns precisely the issue of cutting up the great periods of history (periodisation) and the identification of objects of study validated and perpetuated by current

[9]Cf. my *Humanisme et Islam.*

academic courses for each period so determined. I shall not bring up again the passionate and often irritating as well as sterile debates of the 1970s concerning Orientalism. I shall simply touch upon the narrow-mindedness which still affects research, publications and teaching in the field uniformly designated by the umbrella term 'Islam' with a capital 'I'. Such narrow-mindedness is due first of all to a number of political, social and linguistic forces controlling the ideological manipulations of the Citadel of Islam since the days of independence acquired in the years 1945 to 1960. In the face of such internal upheavals in the vast field of Islam, political sciences as a subject has marginalised classical historicist studies in this field, while at the same time prolonging the ghettoisation of an object of knowledge which has always benefited belatedly from the new developments in the humanities and social sciences applied to Western societies.

It will be easily understood therefore why the challenges of truth in all writing in the field of the humanities and social sciences have informed all my initiatives to inscribe 'applied islamology' in what I call the 'Mediterranean geo-historical space'. Applied islamology is a new discipline differing from classical islamology and also from a way of writing specific to political sciences. It applies cognitive intervention strategies in the current scholarly and ideological debates between Islam, Christianity and Judaism after the appearance of the Qur'ânic fact. Later on, starting from 1800 (Napoleon in Egypt, Great Britain in India and elsewhere), it applies these strategies in those debates concerning faith and reason, tradition and modernity, secularisation and religion, democracy, human rights and religion, development and under-development, terrorism/fundamentalism and Western values, and so on. Such an intervention strategy requires a radical critique of all forms of rationality, or rationalisation, which are the heritage of the long history of Islamic thought. However, care should be taken to confront them with those competing rationalities and rationalisations developed in the Mediterranean historical space. It is in such a space that supportive research surpassing all emotionally charged dialogues must finally examine the upstream and downstream of critical reason. The upstream is the intense, constant, instructive, creative, liberating dialectic, a dialectic which is above all comprehensive of the challenges of truth in historical praxis. The complexity and the fecundity of all that is left to re-examine in this upstream may be read in the following diagram:

$$[\mathrm{WG} \rightarrow\rightarrow\leftarrow\leftarrow \mathrm{PD}] \rightarrow\rightarrow\leftarrow\leftarrow [\mathrm{L} \rightarrow\rightarrow\leftarrow\leftarrow \mathrm{M}]$$

WG: Word of God; PD: Prophetic discourse; L: *Logos*; M: *Mythos*.

Downstream consists of the competitions, the confrontations, the wars, the different forms of violence, false knowledge, obscurantist and alienating beliefs, dogmatic certainties, rigid principles and definitions, desires, passions, emotions…in a word, everything that impels one to fritter away energy, work, resources, and universally esteemed positive effects brought about by men in society. Nowadays, in the more advanced democracies, the ethic of communication advocated by Jürgen Habermas as well as many other new contributions by countless thinkers, researchers, writers, are drowned in the ceaseless noise of the media, everyday concerns, the manipulation of desires, the tyranny of emotions, the turmoil of people's mental universes, the conflicting claims from politics and trade unions… The transmission of knowledge, which is stored up in libraries, reaches only the ever-decreasing fringes of public opinion. This leaves very little time and attention for taking on problems afflicting deprived societies which are subjected to obeisance of the joint mandates of the current powers and so-called religious traditions. The pressure of institutional powerlessness, resignation, omissions and ignorance is the scourge of all contemporary societies, but in particular and more tragically of the more deprived and resourceless societies, once described as underdeveloped in the aftermath of their independence.

These brief remarks help us gauge the state of truth and of the truths in the many cultures which have long since been cohabiting in Euro-American societies. Many questions arise with regard to the political management of pluralism, diversity, identities, of vehemently proclaimed differences, without providing new citizens with the spaces and the cultural resources needed for critical self-evaluation. Semantic disorder, conceptual confusion, and indignation against racism and discrimination get the better once again of the search for clarity, for viable and emancipating values, for the legitimacy which would widen citizens' consensus. While trends towards withdrawal of communities into themselves are denounced, the educational system is not opened to subjects which would make room for the critical expression of deracinated collective memories. In the course of my career as a teacher and researcher I have done my best to open up spaces and ways of transmissions of knowledge and critical confrontations of identities which were affirmed but never truly considered. From Franco-Maghrebin historical developments I have drawn countless examples which support the confrontation of values, truths, the use of reason, overcoming false identities resting on communitarian or nationalist mytho-historiographies, but not even a few swallows make a summer.

I have never stopped reflecting or making others reflect on the travails of truth in its religious pathways and in its modern metamorphoses in Western Europe first, and outside Europe later. Marie-Dominique Chenu found himself in trouble with ecclesiastical hierarchy when he assigned to theology the leading task of 'the intelligence of faith subjected to the test of time' – that is to say to the constant rebuttals that lived history inflicts on the imaginary constructions of reality. Acknowledging that religious truth is a historical one has not been accepted unequivocally in Roman Catholic doctrinal teaching. Islamic thought ignores the historical nature of faith even more, in so much as it has cut itself off from theological and philosophical research since the turn between the thirteenth and fourteenth centuries. This is a historical fact, throwing light on the relationship with truth in the monotheistic pathway after the subversive intervention of the modern epistemological gap. In the many interreligious and intercultural symposia that have taken place since Vatican II this decisive point has more often than not been passed in silence: preference is given to polite and 'tolerant' agreement concerning the articles of faith linked to the syncretic figure of Abraham.

After Marx, Nietzsche and Freud, Foucault has insisted on the necessity of subjecting reason to the tests of political practice and of the conducts of civil society in the face of the so-called rule-of-law state. All the same, it can be observed that there is still a strong tendency everywhere to appeal to Enlightenment reason in order to stigmatise the barbarism of our times. In doing so, the illusion is kept that there exists a form of appeal considered *a priori* an intangible validation authority. Believers proceed in the same fashion by indefinitely commenting the eternal truths of the Holy Scriptures and of what goes under the name of 'living tradition'. In the world-wide appeal to the Enlightenment, which is qualified as new, there is also a tendency to obliterate the teachings specific to the three great founding moments of modern intellectual history as a never-ending quest of the spirit. These moments are:

1. The Enlightenment as envisaged by Kant and eminent thinkers of the eighteenth century;
2. The same Enlightenment readopted and philosophically subverted by the three creative thinkers Marx, Nietzsche and Freud;
3. The great ongoing debates, in the humanities and the social sciences, since the end of colonial empires. There is no question that the sudden crumbling of communist ideology followed by the emergence of a new menace on 11/9/ 2001, has created a new gap, having a before and after in the history of advanced democracies facing the anger of peoples and societies formerly dispossessed of their own history;

All the same, intellectual subversion only concerns the fringe of researchers, thinkers, writers and artists in the West. The members of civil societies in the world's richest countries remain imprisoned in their preferential claim of their consumerism. The historical misadventures of those peoples which are victims of the military and police regimes that have succeeded colonial power, do not inspire any political will aimed at humanising international law and the successful regulation of issues between states. We seem to be content with humanitarian aid, which is also controlled by standing regimes such as those in Sudan, Burma, Zimbabwe, and of course in Palestine. The initiative of allies invading Iraq to eliminate the tyrant has brought about the results we all know of. Whole nations find themselves therefore between the rock of the West and the hard place of predatory states that call themselves national states. They are thus victims of worse spoliations than those they suffered when the customary and cultural codes of traditional societies ruled the roost. Social and political sciences hardly ever consider the issues raised by such historical impasses; they do not even uncover the intellectual and juridical responsibilities of the great decision-makers with regard to the future of the planet. 'Islam' taken hostage by its own believers has become one of the sources of systemic violence propagating itself in the modern world.

Fertile areas for critical thought and the knowledge of contemporary societies have thus found themselves deserted, ignored, often – it is true – because they are inaccessible to researchers of any origin, including native ones. I am thinking for example of the cross-examination by colonisers and colonised peoples of the colonial adventure and of its perpetuation after the wave of independence proclamations. It is a question of abandoning conflicting, polemical and ideological confrontations with their recurrent eruptions of reciprocal condemnations. I am thinking of a 'solidary writing of the history of colonisation' where the two protagonists criss-cross their respective questionings, their subjectivities, the vision of their pasts, presents and futures. The French and the Germans have set a trend in this respect by writing a history textbook used in schools and high schools in both countries. This opens up new and unexpected territories for discovery and the construction of novel truths about the productive conditions of human existence. Such an experience is valid for all countries adjacent to one another, all recurrent wars, all rigid borders, all mutually exclusive systems which to this day still control the setting up of social, state, inter-ethnic and international frontiers.

The work of decolonising minds makes sense both for the dominated and for the dominant powers; it should have started the very day after the independence ceremonies. Instead of this, new stratagems were devised to prolong dependency relationships by turning a blind eye to the behaviour of prevaricating states many of which can only be described as rogue states. Respectable democracies have come to terms with and continue to support such states which turn against their very own peoples. The *Machtpolitik* with regard to the former Third World has had heavy consequences for the destinies of people in Africa and Asia, but did not disturb the conscience of the defenders of human rights. The end of the Cold War has unleashed nationalist movements which had been muzzled up under the Soviet regime; it has fed the expansion of systemic violence on a world scale. The catastrophe of 11/9/2001 should not be understood exclusively as the unleashing of Islamist fanaticism. It highlights the tragedies produced everywhere by the arrogant and triumphant historical march of Western so-called values and the evident regressions of societies twice dispossessed: by the absence of politics from their national 'elites' and by all sorts of pressures by a West happy and proud to announce 'the end of history' and the 'clash of civilisations', legitimising a new discourse of double criteria after the crumbling of the Communist ideology.

The regression of societies subjected to the alienating distortions of an imagined, or even, in some circles, a ghostly Islam, poses with singular force the issue of the reversibility of the 'values' of civilisation. Although intellectually coarse and culturally very limited, the popular Islam of peasants and nomads before the 1950s had a feeling for human dignity, a respect for the person that uprooted, ideologised, migrant, alienated classes of post-colonial societies have lost. On this level of sociological, linguistic and historical analysis of non-Western societies, it may be worthwhile to test the operational and explicative range of what I call the anthropological triangle: 'Violence, Sacred, Truth'.

IV The Anthropological Triangle

The works by René Girard on violence and the sacred are well known. In what follows Michel Serres presents a wide-ranging and original work since it founds an anthropo-philosophy haunted by the issue of the theological essence of Christianity as the only religion liberating humankind from the sacrificial ritual of the scapegoat victim. It will be seen that such a question is important in terms of the complexity of the issue of truth.

> One day historians will ask you to explain the unexplainable: this formidable wave which engulfed our West during the twentieth century. [...] Such abominations are largely beyond the possibilities of historical explanation; in order to try to understand this incomprehensible, what is required is a tragic anthropology on the scale of your anthropology. One day we will understand that this century has extended on an inhumane and world scale your society and individual model. Once again, where does this violence come from? From miming, you said. [...] Now, when everybody desires the same thing, it is a war of everybody against everybody. We can only tell of this heinous jealousy of the same, which opposes doubles and twins transforming them into enemy brothers. Almost divinely performative, envy produces in front of it, indefinitely, its own images made in its image. The three Horaces resemble the Curiaces triplets; the Montagues imitate the Capulets, Saint George and Saint Michael mime the Dragon; the axis of Good acts symmetrically, according to the image, hardly inverted, of the axis of Evil.[10]

In other words, Girard assigns a shared substratum to every human culture: the cathartical lynching which is assuaging and welding the community together. However, he makes use of this anthropological hypothesis in order to give a 'scientific' basis to a theological assertion which has more to do with apologetics than with epistemological validation. It reduces the truth of humankind, of its hermeneutical adventures and of the objective world to Christianity alone, which makes an innocent of the victim and takes a radical stand against all forms of scapegoat sacrifices. It must also be added that there are other researchers, thinkers who have reactivated the Christian exception by a mix of anthropological culture, philosophical criticism and theological discourse, the epistemological status of which is still far removed from that of the humanities and social sciences. A well-known thesis is the one by Marcel Gauchet explaining Christianity as the only religion which has led, historically, to religion abandoning the stage in Western Europe. In the same perspective the Italian philosopher Gianni Vattimo writes:

> This [*the history of salvation as a history of announcement*] may well be the nature of the Judeo-Christian message which is unique in the history of religions and which, aside from all imperialist or Eurocentric claims, gives it a reasonable claim to the status of universal religion.[11]

It will be seen that none of the thinkers and researchers who have at length expressed their views on Judeo-Christianity have said anything about the example of Islam, which – it cannot be denied – has insistently, albeit at times polemically, remained within the long monotheistic tradition. I am thinking in particular of Paul Ricœur and Emmanuel Levinas. This weakens the scholarly thesis as much as the philosophical consistency of their reasoning. Such a shortcoming has led me to widen the anthropo-philosophical space of what I call critical and reflexive anthropo-history. Instead of taking as my starting point the pair violence/sacred alone, I have included truth as well, since the search for it, its postulates, its construction, its contents, its political, speculative and social functions are inseparable and in dialectic tension within the space of a triangle from which it is so difficult to break out, as we are

---

[10]Michel Serres, reply to René Girard, in: *Le Monde*, 16/12/2005.

[11]Vattimo, *Après la chrétienté*, 49, 53.

trapped inside it in spite of the efforts made within the so-called neutral space of secularism.

I have chosen the whole of Sura 9 to show the theoretical pertinence and the informative riches of my heuristic hypothesis on the functioning of the anthropological triangle not only in the case of Islam but for all religions as well as for the modern spaces of secularism. Here I am quoting just two verses in order to make an outline of my approach which can be summarised by three infinitives: to transgress, to remove, to go past. These three verbs refer to spaces of reality, to cognitive strategies of intervention, to disciplinary fields and their methods and thinking tools which are used by emerging reason.

V Brief Illustrations[12]

I shall start from two brief quotations and two Qur'ânic verses to show the tribulations of truth in its diverse dimensions in the contemporary world.

The proclamation of 1/11/1954 assigned the high task of "restoring a sovereign, democratic and social Algerian State within the framework of Islamic principles" to the Algerian war of liberation. In 1959 colonial France at last considered promulgating a law prohibiting the repudiation of women and imposing a judicial procedure for divorce. The *Moujâhid* of 6/7/1955, the FLN daily, refused indignantly such an intrusion into the personal status of Muslims:

> Thus, some Frenchmen, and what is more, Christians or Jews, as is the case, apparently, of Michel Debré [*the then Prime Minister*], have dared to deliberately undermine the Qur'an, which is by its very nature immutable, and to impose France's secular laws on Algeria's Muslims by the sword, and this in the most sacred subject-matter, namely, one's personal status.

Earlier, Bourguiba, who would go on to become the liberator of Tunisian women, had thus spoken in order to refute the arguments of a female militant against the wearing of the veil, in 1929:

> As a result of the regime to which it is subject, Tunisia is fast moving headlong towards the degeneration of its personality [...] In order to fend off this danger, Tunisians must see to it that they safeguard their traditions, which are still the distinguishing traits and therefore the last defences of a national identity in jeopardy.

Here are the two verses which will serve as a basis for our work on the concept of anthropological triangle:

> When the sacred months are over, kill the polytheists [*al-mushrikûn*] wherever you may find them! Capture them, besiege them, ambush them! But if they come to God [*tâbû*] through prayer, by offering the prescribed alms, then let them walk freely, because God is all indulgence and all compassion. (9, 5)

---

[12]Cf. my *Pour sortir des clôtures dogmatiques*.

> The Jews have said: 'Uzayr is son of God';
> The Christians have said: 'the Messiah is son of God';
> These are the words emanating from their lips, just like
> Those pronounced, before them, by the infidels.
> May God annihilate them!
> They are so far removed from Truth! (9, 30)

V-1 The Quotations

In spite of their antiquity, the two quotations continue in 2008 to call forth the same indignation, the same protestations and the same ignorance as when they were first enunciated. Arrogance and ignorance are not just on the part of the colonised; the coloniser is included, displaying the same degree of certitude impregnated with contempt for an obscurantist Islam, a magic mentality oblivious to the contributions of a civilisation of progress. It is the shock of ignorance, tied to unquestionable truths and moral values which are by far superior to those of the enemy. As I write this, the air in Paris is all aflame because of the boycotting of the *Salon des livres* book fair by Arab States and writers. The difference with the confrontations of the 1930s and 1940s, is that the participants are better aware of the history that separates them. But the arguments of the two protagonists obscure the references to this worn-out history served *ad nauseam* in the mytho-ideological versions of official historiography which are still a heavy burden on people's minds. It is a well-known fact that the French parliament has even recommended a form of history explicitly highlighting the advantages of civilisation. The indignation by the great professional historians in France has induced power to step back and to reaffirm the critical independence of historical research and teaching in general, also with regard to such serious confrontations as those arising from colonisation. In Algeria, apart from the repetition of the usual indignation, no need was felt for establishing a shared history of the colonial enterprise as such including colonial practice, starting from a self-critical examination of the two protagonists. Without this both will remain prisoners of official polemics and self-glorification. As a matter of fact, progress in this sense is extremely slow and tragically insufficient.

Therefore, several crucial questions are evaded by the more advanced democracies and post-colonial states which continue to use and abuse the denunciation of 'the cultural and human genocide' to deflect attention from their lack of political and intellectual legitimacy which is evident on both sides. What TRUTH should we look for when systemic political violence, long since denounced in anti-colonial struggles, arises again on a world scale, in civil wars breaking out in the majority of societies deprived of the freedoms and prosperity promised during and after their liberation wars? Social justice, dignity and liberty are on everybody's lips more than truth. This feeds of course on these basic needs of man in society. There can be no truth, nor political legitimacy, in places which see the triumph of systemic violence, of so much frustration and deprivation in populations where young people are more numerous than the over forties. Justice itself cannot come to fruit without a durable peace culture which would put an end to the political programming of conflicts. Even more tragic and ravaging conflicts have been planned by the victors of the second Europe-wide war at a time when colonial empires were still prevalent.

The humanities and social sciences suggest advances towards multiple truths which can be made relatively objective and are respectable as such. Even when reliable truths are identified which should be binding for everybody, it is a persistent fact that political decision-makers do not even think about integrating them into historical action. The logic of political decision-making does not bow under the weight of either the truths purported by the experts, nor under that of ethical values celebrated in illusory or openly deceptive rhetoric, or even less that of the suffering of the defeated. Let us consider the imposing and masterly work of Paul Ricœur, author of, among many other great titles, *Soi même comme un autre*. No sooner had he died than his rich, generous and fertile thought was buried under an avalanche of tragic news, of books and words without a future, linked as they are to mundane polemics and passions.

I could not claim such levels of interest for my own works on the *Critique de la Raison Islamique*. During his life, Paul Ricœur could rely on a wide public in the West reading and commenting his writings. A critical researcher-cum-thinker exploring the extensive field of Islam cannot aspire to such a wide public. After concentrating its efforts on the fight against the communist camp, the West has turned the eradication of Islamist terrorism into a new priority. Nonetheless, the historical road taken by the so-called Muslim world is very different from that of the Soviet Union since the October revolution. The history of colonisation, written single-handedly by capitalist Europe, relieved by the United States after 1945, has been erased from memories reduced to silence – hateful or painful memories, depending on who is speaking. Subsequently the construction of the European Union has opened up avenues for action and hope which have favoured the oblivion of an outcome that was tragic for all protagonists. There are, however, countless truths to be discovered in order to appease so many tragedies which the people concerned passed by in silence.

The nature of these historical truths is such that they have curbed the strategies of geopolitical control of the whole burning region of the Middle East. By deliberately ignoring them, the two protagonists, 'Islam versus the West', have enabled the growth of resentment, mistrust, rejection, the desire for revenge which have led to the tragedy of 11/9/2001 and its aftermath, which was even more dramatic than its previous trials.[13] There is the question of an internal and an external genesis of the recourse to terrorism which should be elucidated as an objective search for truth. I am referring to a search for truth in order to justify the cathartic function of history as an example of what is true and fair.

Before 11/9/2001 there was an objective alliance between national states which 'nationalize' Islam from within, and some powers exploiting the mobilising force

---

[13]For more discussions on this subject, cf. my considerations and analyses in *De Manhattan à Bagdad*.

of radical Islamism in order to extend a better control over future geopolitical and economic divisions of the world. Those who analyse such strategies content themselves with describing such strategic manipulations and gauging their respective chances of failure or success. They hardly ever worry about the disastrous effects of a permanent *Machtpolitik* in the intellectual, scientific, cultural and moral field, in societies tested both by national states unconcerned with their legitimacy and the public good, and by powers supporting or quashing, as the case may be, the murderous violence of subversive movements without viable political programmes.

This is the triumph of a discourse applying double standards: on one hand it calls for compliance with human rights and democratic values while at the same time it perverts or even annihilates the work required by free thought to protect and enrich the critical quest for truth which is inseparable from the quest for meaning, justice and legitimacy in any human society. In this field too, the humanities and the social sciences rely on affidavits in the absence of elementary freedoms, on the cultural, moral and scientific regressions in contemporary Islamic contexts, without elucidating the historical genesis and the ideological, institutional, political, and social factors which have led to the flight of the elites, to resignation and to a feeling of humiliation among people, as well as to the anger of those who no longer wait in this world and decide to join the cohorts of martyrs. There is a need for a great book on the political and psychosocial genesis of the mythology of the martyr since the wars of liberation. All over the former Third World free course has been given to populist religiosity, to regressive ideologies, and to systemic political violence in the sense that it combines internal and external factors. The irresistible expansion of the culture of religious non-belief replaced by the cult of consumerism, of spectacle and profit in the West, has its counterpart within the protagonist 'Islam' in the form of an increase of religiosity bringing salvation, nostalgia for a past grandeur long since forgotten, an uncertain and confused wait for a concrete emancipation at last freed from a dangerous messianic hope. In the current phase of a generalised crisis of reason and of a quest for foundations or for an authority protecting truth, sense and legitimacy which have universal and not just national or communitary range, a quest for a more reliable truth is simulated by finding refuge in the description and deconstruction of ideological or purely imaginary regimes of conflicting 'truths' who continue to offer the basis for the current wars.

In order to open up new fields of investigation, let us remember some of the tensions having educational import between theology and philosophy. This reminder is necessary as theology, which is extremely marginalised, continues to consider possible ways of renewing itself by turning for help, like philosophy, to new questions, methodologies and conceptualisations not just in the humanities and social sciences, but also in the revolutionary progress of the life sciences. I would stress the fact that such a theology, open to the great epistemological debate in the neurosciences, life sciences, the status of the foetus and of the virtual human person, not to mention the recent debates unleashed by history, linguistics and anthropology, is practised within Christianity (Protestantism and Catholicism), in a more limited fashion within Judaism, and very exceptionally within Islam. This means that the medieval concept of 'true religion', to the exclusion of all others, continues to prevail in inter-religious dialogues and more dogmatically in the ritualistic expressions of religious belief. In spite of its important overtures to other religions after Vatican II, Catholic doctrine continues to defend the idea of a hierarchy of religions in relation with their respective truth regimes. One should get rid of all obsession with truth, knowing that the most ancient truths, the most rooted and most recurrent ones in societies (I am thinking of the current vogue for religions), are nothing but beliefs which are victorious while eliminating other established, concurrent or new beliefs.

The critical activity of reason in its different fields (in terms of religious, theological, philosophical, political, economic, dialectic, juridical, and ethical reason) does not aim so much at discriminating the rational from the irrational, the imaginary or the ideological within each historical path linked to each logosphere; it must move across all frontiers in order to explore and compare the whole of thought and belief systems which base themselves on the Truth or truths which have not moved across the test of comparison and of confrontation; not just with the universal, always invoked though never realised, but with that 'which can be made universal' in such a way that it maintains everywhere the dynamic processes of dialectics, of confrontation and of validation. Monotheist religions which during the Middle Ages experienced a demanding confrontation with philosophical reason continue to mistrust a philosophical attitude and above all to exclude each other from the status of 'true religion' as a candidate for the universal. What has been called the return to religion in the West exacerbates the quest for identity and the claims of a right to differ by widening the mutual exclusions nourished over the centuries by the three monotheist versions of election, of the Alliance, of the privilege conferred by the role of unique possessor of the true revelation of the Word of God. Modern reason introduced an alternative to such exclusionism: it subjects to its critical scrutiny all truth regimes, but it tends to monopolise the production of a universal truth equally shared by human beings. From thence derives the expansion of the culture of non-belief which refers back both to the relativity and the contingency of any truth, and to categorical imperatives for all production, reception and practice of truth.

In all the cases in which truth is invoked or questioned, a major debate is created which may even turn into conflict. All true knowledge or values cannot but be conflictual. The priority given by Kant to the preliminary study of 'the conditions that make a pure reason and a practical reason possible' while always aiming at consensual truth, maintains a certain methodological usefulness but it does not put an end to conflictual truth regimes. It is well known that Marx produced a critique of political economy which was extended, completed and made more profound by Nietzsche in his genealogy of values. The first critique was appropriated and obliterated by the so-called proletarian revolution; the second was ill-understood, abandoned and above all diverted from its philosophical ambitions by the return of belief and non-belief systems under the sign of empiricism, of individualistic pragmatism, of savage liberalism, of the marketing of culture and thought, of the World Trade Organisation… A significant initiative – in terms of new conditions in which to exercise critical thought – has just been taken by the newspaper *Le Monde*, which has been republishing the great classical works of philosophy, at prices within the reach of the wider public that had been kept on the margins of the debate on the liberating intellectual and artistic combats even in those societies where all children have access to schooling and all citizens have, in principle, the right to knowledge and culture. This is a gauge of the persistence of obstacles in the way of the production and advances of truth in human societies where the weight of the social construction of truth is still predominant.

A problem arises here, highlighted by political scientists speaking of the end of Western monopoly to direct the forces that produce human history. This consideration replaces the theory of 'the end of history' announced by Francis Fukuyama at the same time as the 'clash of civilisations' at the beginning of the 1990s. If the rise of India and China is the harbinger of a multipolar world in the area of production and of economic and monetary exchanges, the functions of reason and of humanistic culture oriented towards a more effective and widespread emancipation of the human condition, do not follow the pace and the effectiveness required by the entry into competition of cultures that are separated by wide gaps or can even be considered regressive if compared to the spectacular advances in scientific research and technological innovation. Without a doubt such advances produce 'disposable thought' in societies of consumerism, of spectacle, of huge profits and social exclusion. Such a situation increases the intellectual, scientific, cultural and spiritual responsibilities of a West whose supremacy since the nineteenth century has to a great extent contributed to the deprivation of the rest of the world in crucial areas where the majority social actors are not enjoying the full and guaranteed status of democratic citizenship.

I shall now go back to the example of Islam. I have been wondering for a long time about the forces of dislocation operating on cultural traditions under the colonial domination. Only fragments of modernity were made accessible to a minority of the happy few, while the majority of the population had to survive with local, limited and outdated values and collective memories. Oral cultures expressed in local dialects, ritualistic Islam mixed with archaic beliefs were offering very limited useful knowledge which could help to cope with the historical challenges of modernity. This cultural, social and intellectual gap generated the inevitable 'nationalist' ideology which led to the wars for liberation. These wars hastened the disintegration of the traditional archaic society, while the so-called historical leaders who conducted the wars were as ignorant of what should be the political future of the societies as the large majority of militants enrolled in the struggle. They were equally cut off from the collective memories with their respective temporalities; they shared the same ignorance of the conditions required by that high task of 'restoring a sovereign, democratic and social Algerian State within the framework of Islamic principles'. This holds true for many other countries, too. We know how the 'principles of Islam' tuned out to political slogans were used by fundamentalist defenders of the Islamic Muslim polity. They recovered the authentic 'Medina Model' as opposed to the reliable alternative to the 'Western Model'.

The party states that took over from colonial regimes have all improvised 're-personalisation' policies (communitarianisms and a quest for 'murderous identities') which have made the regression process worse. They created a crisis of values, and what's more, of functions of free critical reason. The civil wars still devastating so many peoples started the day after independence proclamations and are still rife in several countries to this very day. I shall say in lapidary fashion that the catastrophe of 11/9/2001 must be thought of in the perspective of a 'greater reason' taking on itself the onus of its omissions, its abdication of responsibility, its choices of violent domination and exploitation in Europe; its oblivion, its breaches, its considerable delays, its mytho-ideological wanderings around the world at large twice dispossessed between 1800 and 2008: dispossession by colonisation, and afterwards, from 1945 to 2008, by national 'elites' which have 'monopolised the exercise of legal violence' according to the general definition of the state by Max Weber.

What must one understand by 'greater reason'? Greater than which other reasons serving as the basis for systems of thought and civilisation, 'values' in bloody competition since 1945? I have often referred to 'emerging reason' also, to name all "the fights engaged in by the spirit for the spirit with the resources of the spirit"[14] in all known traditions of thought, restored more or less adequately by historians and identified by anthropologists throughout the world. Although European erudition has accumulated an impressive production on non-Western cultures, there is still much to do in order to create conditions for mutual creativity, for the interpenetration of systems of thought and the works of culture and civilisation. As a result of migratory fluxes throughout the world, the construction in the European Union of a space of new citizenship, we are entering the historical phase of overstepping and transgressing frontiers which are jealously guarded by sovereign states and clerics busying themselves with the cult of 'true' religions as well as identities closed by virtue of their self-celebration. It will be necessary to leave behind very quickly the traps and prisons of *Jihâd versus McWorld*, as in the book title by Benjamin Barber.

V-2 The Two Verses

How should one read these two verses? There are several reading protocols of the great founding textcorpora of religions. I cannot go into details here.[15] I shall mention the reading protocol of the believer, the philological and historicist reading, the literal or esoteric reading, the grammatical and lexicographical reading, and the theological

---

[14] This phrase is the translation of a celebrated confession by 'Abd-al-Qâdir al-Jîlânî, hanbalite mystical theologian who became one of the greatest saints of Islam (d. 561/1166). In Arabic, he uses the term *haqq*, a concept expressed in the Qur'an and notably enriched by the great mystics of the classical period. *Haqq* refers at the same time to truth, justice, what is right, the actually real. Here is the quotation, deserving a long commentary, on the truth experienced, interiorised in the personal experience of the divine: *Nâza'tu-l-Haqqa bi-l-Haqqi li-l-Haqqi*. I translate '*haqq*' by 'spirit' because the search for the *haqq* in the mystic engages all the faculties of the spirit, reason, intelligence, memory, imagination and capacity to contemplate the imaginary. Important elaborations on the purport of *Haqq* can be seen in the work of the same author entitled: *al-ghunya li-tâlibi tarîq al-Haqq*, Cairo 1304.

[15] Cf. my *Pour sortir des clôtures dogmatiques*.

or juridical reading. Each of these protocols leads to more or less reductive 'truth regimes', to partial or even false knowledge, to anachronistic projections on fragments which are decontextualised and totally indifferent to the writing conditions of critical history, to synchronic and diachronic linguistics, to the movement from oral culture to written culture, from the first enunciation by the mediator Mohammed Ibn 'Abdallah called God's prophet or Messenger, in the enunciation that he has passed on.

Such ignorance, such omissions, such manipulations which are shared by interpreting communities subjected to the traditional theology of Revelation are undertaken and scrutinised by the reading protocol appropriate to 'emerging reason'. The complexity of this protocol shall appear in the advances of analysis and critique.

Read in their literal meanings, the verses 9,5 and 9,30 will delight all those who reduce Islam to the conquering religion holding people prisoner in the dilemma: conversion or death. It is a solid base for all islamophobes, today's and yesterday's. Nowadays peaceful believers are embarrassed by the explicit appeal to kill all polytheists refusing conversion to the 'true religion'. Since they cannot deny the evidence of the commandment, many avoid citing such an injunction, especially when it is known that medieval commentators call it with pious zeal "the verse of the sabre (*ayat al-sayf*)". On the other hand, fundamentalists renew the call of this verse with the zeal of the ancients, and brandish it with fervour in order to legitimise the 'just war' (*jihad*) against the known 'aggressors' of Islam. They torture the text and appeal to pacifist verses to demonstrate that Islam is a religion of peace and tolerance. The just war is a canonical obligation for believers only in the case of legitimate defence or of breach of alliance between clans and tribes. This breach is precisely what the Sura denounces from the start of the first verse: "Allah and his messenger disengage themselves from a pact not observed by their partners" (*bara'atun*), which is what opens up the possibility of a return to hostilities.

The philological and historicist protocol spends more time on the history of words, the etymologies, the influences and the transformations of meanings. It will signal false meanings, the misinterpretations of the literalist who totally ignores the dangers of anachronisms. It will disdain the fantastic interpretations resting on legends or mythological tales which were reported as historically true. Philology has long since been the queen of the disciplines used in the historico-critical method and the critical editions of ancient texts. A typology of the protocols can thus be drawn up to mark the sudden changes and the new contributions of that which is practised under the name of discourse criticism in its different written enunciations or articulations, without going by the way of oral expression.

Whether it be the Bible, the Gospels or the Qur'an, when reading a verse or of a longer textual unit three levels of contextualisation must be kept in mind: (1) the 'discourse situation' of the first oral enunciation; (2) the immediate textual context where the verse or fragment has been inserted; (3) the context incorporating the Fixed Official Corpus: the whole of the tales, the fragments and large units collected and divided up in the volume called Bible, Gospel or Qur'an. A large number of problems can now be detected which were ignored for centuries by all the reading protocols mentioned above.

I shall not follow all these stages in such a brief contribution: it is essential that the reader's attention should focus on the great truth challenges present in each protocol. In the case of the Qur'an, it is certain that the statalisation of Islam and the appearance of opposition movements such as the Islamic Salvation Front (FIS) in Algeria, the Talibans in Afghanistan, Hizbollah in the Lebanon, Hamas in Palestine… have seen a moving away from an exegesis, framed within a long repetitive tradition, to a 'savage' exegesis escaping all control and clinging to militant pressures entirely devoted to the imperatives of combat ideology. For example, in verse 9, 30 mention is made of a certain 'Uzayr recognised as son of God by the Jews. The ancient exegetes had the merit of questioning the identity of this 'Uzayr. Tabarî stresses that it is unusual for Jews to express such a belief. Modern erudition suggests that 'Uzayr was in fact the High Priest Ezra, who had contributed to the reconstruction of the Temple after the return from Babylon to Jerusalem. This historical curiosity no longer haunts the vast majority of Muslims since any historical reading of the verses questions the orthodox representation of a Word of God descending on earth and communicated word by word by the prophet. Here again is a very illuminating example of the conflicts of truths with regard to names, roles, events, or polemical positions which have been turned into untouchable dogmas.

Let us add a few remarks on the meaning of the links between the two quotations with regard to contemporary actors and the uses made of the Qur'ânic text. Firstly, let us comment on the links between the two quotations and the appeal or absence of appeal to the 'immutable Qur'an'. The outraged text of the *al-Mujâhid* newspaper puts the stress on the scandal of *Christians* claiming the authority, under the colonial regime, to modify measures of the Divine Law concerning the status of women. Bourguiba links the emancipation of women (these are the 1930s) to the danger of degeneration of Tunisian personality. Women are the certain and faithful guardians of customs perpetuating the distinguishing traits of national identity. The immediate emotional adherence to this vocabulary of 'values' erases in people's minds the violence implicitly experienced by women without ever making it explicitly known until very recent times. Moreover, the evocation of the endangering of national identity by the colonial occupier evokes the need for a close combat, without making clear what price was to be paid over time in order to maintain an archaic and backward patriarchal regime after the liberation and installation of a national state. Family law and succession law continue to be in force in several countries, thanks to the undisputable argument that human legislation cannot alter explicit orders of the Word of God. I know no Muslim jurist who would open up the ways of a theology establishing the cognitive status of Qur'ânic discourse as a whole. Personally I have been doing this since the 1970s, though my intervention was closely linked to that of non-Muslim orientalists. In this way the structure of shared image of Muslim belief fixed since the fourth/tenth centuries continues to control and curb all initiatives of intellectual and cultural liberation by opening, as I have done, the site of a critique of juridical reason in contemporary Islam.[16]

Critical thought, both political and juridical, is blocked at the same time by traditional jurists, guardians of the Sunni and Shia orthodoxies and also by the post-colonial states which statalised religion by the bureaucratisation of the control of theology,

[16]Cf. my *Humanisme et Islam.*

politics and law. The quotation of one verse or authentic *hadith* is enough to stop all free debate on issues concerning personal status (*ahwal shaksiyya*). In other words, the activity of critical scientific reason is under the strict dominance of a shared image of belief. Everyday social and political discourse suffer under the perverse and often paralysing effects of the emotional register of belief combined with that of a vehement nationalism continuously set aflame by the intolerable arbitrary politics of 'the West' in the whole of the geopolitical space renamed the 'Greater Middle East' under the reign of the Bush administration. Thus, the explosive and perverse combination of the political and of the religious mental universe has considerably enlarged the fields of taboo subjects and vocabulary within the whole of what is wrongly called the 'Muslim world'. The issue of truth is the first great taboo, since there is no truth that can be articulated in the Qur'an or enunciated by the human spirit without validating it through Qur'ânic and prophetic teaching. This, at least, is the received belief concerning the Word of God as consigned in the corpus called *Mushaf.*

Sergio Romano:

Thank you very much, Professor Arkoun, you have very effectively conveyed your message. So far I have not mentioned the Jewish religion. Let me just invite Geza Vermes, who has crossed the border between two religions. He was born in Hungary. He became a Catholic priest and then he joined the Jewish faith, becoming one of the greatest scholars in the field of Early Christian religious studies, including the study of the Dead Sea Scrolls. He is a Fellow of the British Academy and of the European Academy of Arts, Science and Humanities.

Geza Vermes:

Thank you, Mr Chairman. I think I ought to preface my talk by saying that my religious odyssey, or the absence of it, has nothing to do with the paper that I am going to read. It will be simple, down-to-earth, and something in the real world.

The Truth About the Historical Jesus

Geza Vermes:

There are two kinds of TRUTH about Jesus Christ. The first is the *Gospel truth.* Its veracity is vouchsafed by faith. In the believer's eyes no contradictions do, or even can, exist in the divinely inspired Gospels. Appearances to the contrary should be ignored or reconciled. For instance the Gospel of John gives a historically acceptable account of the condemnation of Jesus: he was arrested a day before Passover and, without the mention of a Passover meal and a formal Jewish court process, he was brought before Pilate, accused of being a revolutionary and sentenced to crucifixion. In the other Gospels, in a historically unlikely fashion, the arrest of Jesus followed by a trial by the Jewish Sanhedrin on the charge of blasphemy, took place after the Passover meal (the Last Supper), and Jesus was pronounced guilty on the night of the feast itself. Yet, no believing Christian asks how the supreme tribunal of Judaea could try a capital case during one of the major festivals, or more simply, how the two stories hang together.

The second kind of truth is less certain than faith and is approximated since the age of the Enlightenment by means of 'scientific' historical inquiry. This quest strives to discover the TRUTH, but succeeds to retrieve only morsels of it. The historian's task is to assemble a monumental jigsaw puzzle of which many parts are still missing. My catching title promises more than anyone can deliver. A more modest 'Towards the truth about the historical Jesus' would be closer to what will follow.

Until the mid-eighteenth century *Gospel truth* wholly dominated the Christian world and it has continued to do so in conservative ecclesiastical circles up to the present day. The certainty did not result from the blinding effect faith exerted on the historical evidence. As early as the second century, divergences among the New Testament records were noted by perspicacious Church fathers and a deliberate attempt was made to harmonize them, producing the so-called *Diatessaron*, the four Gospels in one, but after some initial success the innovation failed and the traditional four Gospels survived. Thus later Church fathers were perfectly aware that the two genealogies of Jesus in Matthew and in Luke were incompatible, but they launched the seemingly brilliant idea that Matthew traced the ancestry of Jesus through Joseph, while Luke did so through Mary. They turned a blind eye to the fact that among Jews a genealogy was expected to follow the male line.

Quest for the human figure of Jesus began with Samuel Reimarus (1694–1768) in mid-eighteenth century and has characterized academic Gospel criticism up to the present day. For the first two hundred years it was essentially a German academic pursuit, though from the late nineteenth century onwards with a smattering of British, French and American contributions. It aimed at the rediscovery of the *historical Jesus* and sought to distinguish him from the *Christ of faith*. Its initial stage ended with the anticlimactic *Geschichte des Leben Jesu Forschung* (*Quest of the Historical Jesus*) by Albert Schweitzer (1875–1965), who in 1906 qualified the whole process far too subjective to be worthy of continuation. According to Schweitzer, each scholar produced a Jesus in his own image and resemblance.

From the 1920s to the 1950s historical Jesus research became rather unfashionable under the influence of the great German scholar, Rudolf Bultmann and his new literary-critical school of *Formgeschichte* or form criticism. In 1926, he advanced the memorable statement which de facto excommunicated Life of Jesus inquiry in the wide academic circles over which he ruled: "We can know almost nothing about the life and personality of Jesus since the early Christian sources show no interest in either".[1] For Bultmann the setting of the Gospel message was not the life of Jesus; the evangelists were catering for the needs of the nascent church. After

---

[1]Bultmann, *Jesus and the Word*, 14.

a thirty year long silence, the historical interest was slowly rekindled in Germany; it was short-lived and without noteworthy results.

In the 1970s, for the first time in two centuries, the main scene of activity left Germany. It first moved to England, the first three main contributions were produced in Oxford, and soon after to the United States. The principal emphasis lay, not on the Hellenistic background of the early church as in form criticism, but on the Jewishness of Jesus in the wake of the discovery of the Dead Sea Scrolls, the renewal of research in postbiblical Judaism and in the first century AD Jewish historian Flavius Josephus. The trend is clearly shown by the new titles: *Jesus the Jew* (1973), *Jesus and Judaism* (1985), *The Historical Jesus: The Life of a Mediterranean Jewish Peasant* (1991); *A Marginal Jew: Rethinking the Historical Jesus* (1991–2001) and *Jesus of Nazareth, King of the Jews: A Jewish Life and the Emergence of Christianity* (1999). While in the scholarly field the Jewish Jesus had become the dominant figure for all researchers, with or without religious belief, for the numerous Roman Catholic practitioners of the quest, the whole issue was put back in the melting pot in 2007 by His Holiness Benedict XVI. In a best-selling book, *Jesus of Nazareth*, published under the alias of Joseph Ratzinger, the Pope declares that the Gospels' Christ of faith *is* the historical Jesus, thus turning the clock back by several centuries. Pope Benedict bravely invites fellow scholars to contradict him if they so feel inclined, but the big question is whether Catholic biblical experts will have the courage to join Ratzinger's independent critics.

Now let's face the main issue. The student investigating the problem of the Historical Jesus is confronted with a concatenation of difficulties. Everybody except the desperately naive knows that the Gospel sources are not strictly historical and postdate the events by decades. The earlier letters of St Paul won't help as their author never knew, or showed interest in, the Jesus of flesh and blood. The four Gospels written some fifteen to fifty-five years after Paul, in the form of biographies, formulate Jesus' teaching adapted for the needs of the early church. Moreover, their readers had a Greek linguistic, and a Graeco-Roman cultural background, yet they were to receive a Jewish religious message originally formulated in Aramaic. We are facing, as I will show, the *traduttore traditore* syndrome. Finally let's not overlook the subjective element. Believing Christian scholars will be preconditioned by their faith and will suspect non-believers of harbouring anti-Christian vested interest. In the circumstances, it is hard for someone who is not influenced by faith to proceed and, what is more, appear to proceed without fear or favour, *sine ira et studio*. O Tacitus, please help!

The historical Jesus can be retrieved only within the context of first century Galilean Judaism. The Gospel image must therefore be inserted into the historical canvas of first century AD Palestine with the help of the works of Flavius Josephus, the Dead Sea Scrolls and early rabbinic literature. It must be borne in mind that Galilee in the age of the New Testament was a relatively independent territory ruled between 4 BC and AD 39 by Antipas, one of Herod's sons. For instance, in the Galilean tetrarchy, unlike in Judaea directly administered by a Roman governor, the tribute collected by the publicans or tax officials went to Antipas, and not to Augustus or Tiberius. Rome's military presence was not directly felt. The centurion, mentioned in the Gospel, was no doubt a veteran settled in Capernaum, and not the commander of a local garrison.

Against this background what kind of picture of Jesus emerges from the Gospels? That of a rural holy man, initially a follower of the movement of repentance launched by another holy man, John the Baptist. In the hamlets and villages, rather than in the cities, of Lower Galilee and the lakeside, Jesus set out to preach the coming of the kingdom of God within the life time of his generation, and outlined the religious duties his simple listeners were to perform to prepare themselves for the great event. An eloquent popular preacher, Jesus manifested his spiritual power by exorcisms and healing. His audience remarked that "he taught with authority", namely, curing the sick and liberating the possessed, and "not as the scribes" who could only quote the Bible to prove their sayings. His cures consisted in faith-healing: they required trust on the part of the sick. He invited them to believe in his healing power as man of God. Indeed, he went so far as to identify this faith as the cause of the recovery: "Your faith has made you well", he reassured a sick woman (Mk 5:34).

In behaving as he did, Jesus conformed to a pattern of charismatic behaviour attested among Jews throughout the ages and down to his own time. The biblical prophets Elisha, Elijah and Isaiah are credited with miraculous healings and resuscitations. Similar phenomena are ascribed in rabbinic literature to holy men living in the age close to the New Testament. Honi in the first century BC and the Galilean Hanina ben Dosa in the first century AD were renowned for their miraculous rain-making power; Hanina's fame also comprised healing, including healing from a distance like Jesus, and general wonderworking. Flavius Josephus (AD 37-c. 100) reports not only on thaumaturgists of Old Testament vintage, like the prophet Elisha, but explicitly mentions Honi whose wondrous intervention ended a disastrous drought shortly before Pompey's capture of Jerusalem in 63 BC. He also refers to Jesus in the days of Pontius Pilate (AD 26–36) and calls him a "wise man and performer of astonishing or paradoxical deeds". The reliability of Josephus' notice about Jesus[2] was rejected by many in the nineteenth and early twentieth century, but it has been judged partly genuine and partly falsified by the majority of more recent critics. The Jesus portrait of Josephus, drawn by an uninvolved witness, stands half-way between the fully sympathetic picture of early Christianity and the wholly antipathetic image of the magician of Talmudic and post-Talmudic Jewish literature. "Wise man" and "performer of paradoxical deeds" are genuinely Josephan phrases which no Christian interpolator would have found potent enough to describe the divinised Christ of the later church.

The contour of the historical Jesus, lifted from the Synoptic Gospels, suggests a magnetic prophetic figure who was convinced that the aim of his mission was to bring his repentant Jewish followers into God's new realm. This kingdom of heaven was foreseen in many of Jesus' parables as the outcome of a quiet and imperceptible change rather than a cataclysmic transformation in the not too distant future.

[2]Flavius Josephus, *Antiquities of the Jews* (commonly abbreviated as Ant.), 18: 63–64.

It would seem that according to the evangelists Jesus considered himself, and his well-disposed contemporaries depicted him, along such prophetic-charismatic lines. Jesus for example explains his rejection by his family and fellow citizens of Nazareth by the well-known saying that at home no one is recognized as a prophet. He was also regularly alluded to by non-local contemporaries as the great prophet from Nazareth. In the anecdote of Caesarea Philippi, Peter's answer to Jesus' question, "Who do men say that I am?", follows a similar turn. Jesus, Peter said, was believed to be a prophet, or the returning Elijah or John the Baptist revived. But when pressed to reveal what the circle of disciples thought of Jesus, Peter confessed, according to Mark, that he was the Messiah, or, according to Matthew, the Messiah, with the added synonym, "the Son of the living God". The latter phrase was understood in Gentile-Christian theology as a move towards the recognition of the divine status of Jesus.

In the course of my research which led to the writing of *Jesus the Jew*, it was impossible not to notice that church tradition tended to attribute the maximum of significance to the honorific titles applied to Jesus by the evangelists. I decided therefore to set up a quasi scientific experiment. I said to myself: Let's try to establish the correlation between the features of the Jesus portrait of the Gospels and meaning of the designations such as 'Messiah', 'Lord', and 'Son of God' in the mind of the contemporaries of Jesus. To achieve this, we must forget the Greek understanding of the terms by the Gentile readers of the Gospel; get rid of 2000 years of superimposed Christian interpretation of the New Testament, and switch instead the searchlight on Jesus' Aramaic-speaking Jewish audience on the shore of the Lake of Galilee. What was the original meaning of the message and what did the original addressees make of it?

To start with 'the Messiah', the Greek *Christos*, if a pollster had interrogated the men in the street in Palestine two millennia ago, asking for a definition of 'Messiah', he would have heard people mumbling about the greatest Jewish king, who would defeat the Romans, while the more religiously minded would have added that the Messiah would be just and holy, and would subject all the nations to Israel and to God. In more peripheral circles, such as the Dead Sea sect, several Messiahs were expected, one royal, one priestly, and possibly one prophetic. But even the don't knows would have had an idea about the messianic age, filled chock-a-block with miraculous events. According to the words put into the mouth of Jesus, this would be the time when "The blind receive their sight, and the lame walk, lepers are cleansed and the deaf hear ..." (Mt 11:5). Did Jesus present himself or did the evangelists portray him as a warlike royal pretender? The answer must be no. Jesus always forbad his disciples to proclaim him the Messiah, and when confronted with the question, "Are you the Christ?", his regular reply was evasively negative: "That's what *you* call me", he kept on saying, not I. By contrast, the non-bellicose wonderworking figure standing in the shadow of the messianic age fits him perfectly. It tallies with the picture of the Galilean healer, exorcist and preacher so prominent in the Gospels of Mark, Matthew and Luke. In his answer to the question of John the Baptist whether he was the one who was to come, Jesus simply pointed to the events surrounding him: the blind see, the deaf hear, the lame walk, the lepers are healed (Mt 11: Lk 7:22).

The title 'Lord', *Kyrios* in Greek, carried high associations in the first century. It pointed to the emperor, the Lord Caesar, whose Latin epithet was divine as in *divus Augustus*. In turn among Greek-speaking Jews, whose Bible the early church appropriated, *Kyrios* (Lord) was the regular substitute for the Hebrew four-lettered sacred and secret name of God. Quite naturally, in the Gospel read in the Greek churches the Lord Christ (*Kyrios Christos*), promptly acquired divine flavour. By contrast, in Jewish circles, with an infinite gap between the divine and the human reality, such a combination was well-nigh inconceivable. Beside Caesar and God, what other meanings did the title 'Lord' possess? What did the Galileans imply when they address Jesus as 'Lord' or *Mar* in Aramaic? The title, reminiscent of 'Sir' in English, could refer to a variety of persons: to a secular dignitary, to the head of the family, to an authoritative teacher, to a prophet and to a miracle worker. The last three nuances perfectly suit the Jesus portrait of the Synoptic Gospels.

Finally the appellation 'Son of God', the title in the Hellenistic world of the deified Roman emperor and synonymous with God in early Christianity, is nowhere attested in that sense in Judaism. It is, however, capable of carrying at least five other meanings. It can designate an angel in the superhuman world. In the terrestrial domain, each Jew was entitled to call himself 'son of God'. But the term underwent a series of restrictive interpretations. In the post-exilic age only the Jews whose heart was circumcised and filled with holy spirit were allotted that name. Also both the Bible and the Dead Sea Scrolls assign filial status to the Messiah, metaphorically the son of the living God. Moreover, some charismatic contemporaries of Jesus were referred to as sons of God. For example, Honi, who managed to produce rain by pestering God, was compared to a son importuning his long-suffering and loving father. Finally, there is the image of the divine voice from heaven proclaiming someone the 'son of God'. This is reported about the Galilean Hanina ben Dosa. When he had put an end to a persisting drought that threatened the survival of the nation, a divine testimony was heard proclaiming: "The whole world is sustained by Hanina, *my son*." This episode closely resembles the heavenly utterance recorded about Jesus at the moment of his baptism: "You are my beloved son in whom I am well pleased" (Mk 1:11). Both sayings indicate that in Jewish parlance 'son of God' implies divine favour rather than the sharing of the divine nature.

To recapitulate, the philological, literary and historical analysis of the Semitic meaning of Jesus' titles corroborates his image as it emerges from the Synoptic Gospels. Hence the only reasonable conclusion to draw from a combined study of the Gospel picture and the honorific titles is that the historical Jesus was a Galilean charismatic whose aim was to conduct his repentant Palestinian Jewish contemporaries into the spiritual realm called the kingdom of God through preaching, healing and exorcizing. Traditional Christianity does not stop at this portrait of the human Jesus, but overlays it with the majestic image of the Christ of faith arising from the mystical meditations of Paul and John and the Hellenistic philosophy of the Greek Church Fathers.

Time constraints preclude the presentation of a methodology that would enable the scholar to sketch Jesus' authentic Gospel. We may deal with it in the forthcoming discussion. In a nutshell, Jesus' preaching was centred on God, the heavenly Father, on the dignity of all human beings as children of God, on life turned into worship by total trust, on an overwhelming sense of urgency to do one's duty without procrastination, on the sanctification of the here and now, and above all, on the love of God through the love of one's neighbour.

To conclude, because of the cross, the task of Jesus remained unfinished. Yet despite the apparent failure of his mission, his magnetic impact was so profound that, instead of abandoning the cause, his disciples began to look forward to his imminent second coming. When by the mid-second century Jesus failed to return, Jewish Christianity progressively faded away, while St Paul's Gentile church survived and after Constantine set out to flourish ... albeit in an alien form that would have nonplussed Jesus the Jew.

Panel and General Discussion

Sergio Romano:

Thank you very much, Professor Vermes, for this fascinating historical analysis. We should begin the discussion with Professor Jan Assmann, a German Egyptologist. He has taught at Munich, Heidelberg, Paris, Göttingen, at the Ecole Pratique des Hautes Etudes in Paris and a number of American universities. He is at present pursuing excavations on the western bank of the Nile at Thebes.

Jan Assmann:

Being an Egyptologist, I look at these theological questions from the outside and, I must confess, from a very distant and rather secular standpoint. I would like to start with reading to you a passage from Moses Mendelssohn's book on Jerusalem and religious power published in 1783, 225 years ago. There he writes:

> I believe that Judaism knows nothing of revered religion in the sense in which it is taken by Christians. The Israelites have a divine legislation: laws, commandments, statutes, rules of life, instruction in the will of God, and lessons how to conduct themselves in order to attain both temporal and spiritual happiness: those laws, commandments etc., were revealed to them through Moses, in a miraculous and supernatural manner, but no dogmas, no saving truths, no general self-evident propositions. Those the Lord always reveals to us, the same as to the rest of mankind, by nature, and by events, but never in spoken or written words.

And then he continues, distinguishing between three types of truth. First,

> Religious dogmas and propositions of immutable truths of God, of his government, of providence, without which men can neither be enlightened nor happy. These were not forced on the belief of the people, by threats of eternal or temporary punishment, but, suitably to the nature and evidence of immutable truths, recommended for rational consideration. They needed not to be suggested by direct revelation, or promulgated by words or writing, which are understood only in this or that place, at this or that time. The Supreme Being revealed them all to all rational beings by events and by ideas, and inscribed them in their soul, in a character legible and intelligible at all times, and in all places.

Secondly,

> Historical truths, or accounts of occurrences of the primitive world, especially memoirs of the lives of the first ancestors of the nation; of the covenant which God entered into with them, and his frequent promise to make their descendants a nation dedicated to himself.

These "historical truths" – though this corresponds of course to what Professor Vermes calls 'Gospel truth' and not historical truth –

> contain the groundwork of the national union, and, as historical truths, they cannot, according to their nature, be received otherwise than on trust; authority alone gives them the necessary evidence.

So, this corresponds to 'Gospel truth', but Moses Mendelssohn calls them 'historical truths' because of their relationship to time. Thirdly,

> Laws, judgments, commandments, rules of life which were to be peculiar to the Jewish nation, and by observing which, it was to arrive at national, as well as every single member thereof, at individual happiness. The lawgiver was God himself; God, not in his revelations as Creator and Preserver of the universe, but God as Lord Protector and ally of their forefathers, as the liberator, founder, and leader, as the king and ruler of that people.
>
> These laws were revealed, that is, they were made known by the Lord, by words and in writing. Still, only the most essential part thereof was entrusted to letters; and without the unwritten laws, without explanations, limitations and more particularly definitions, even these written laws are mostly unintelligible, and must become so in the course of time, since neither any words nor written characters, whatever retain their meaning unaltered, for the natural age of man.

So, Mendelssohn reserves the concept of revelation and historical truths to the last type of truth: the specific rules of Jewish life. Besides the irreducible plurality of truth, like this one, the truth of concrete religions, there is one general human religion based on the natural revelation of immutable truth common to all rational beings. Mendelssohn invites us to distinguish between, on one hand, local truths that assert their binding authority within the confines of a given concrete religion where they achieve the status of revelation, and, on the other hand, universal truths, which are never revealed, let alone written down, but only adumbrated and to be aimed at in discursive approximation. Last but not least, Mendelssohn arrives at a kind of cultural or religious relativism, stressing that religions only exist in the plural, without, however, giving up the idea of universal truth altogether by distinguishing two levels of truth and by assigning the concept of revelation to the lower level of local truth. I think that this two-level theory of truth is a good example of what Dominique Schnapper has called 'relative relativism', liberating us from the pitfalls of absolute relativism.

Now I would like to comment on the paper of Professor Vermes. I totally agree that distinguishing between truth of faith and historical truth is a good thing, and I also subscribe to Professor Vermes' construction of what can be confirmed as historical truth by a contextual approach, taking the Koran and Talmudic authors as well as Josephus into account. What I would like to comment upon is the meaning and genesis of the concept of truth, of faith – *Glaubenswahrheit* in German – and the concept of faith – *Glaube, foi*, faith or belief, *pistis* in Greek and *emouna* in Hebrew – in general.

What does 'faith' mean? As an Egyptologist, my first step in such matters is always to try to translate those words into Egyptian. In the case of faith, this attempt is doomed to failure. There is nothing, either in mythology or the lexicon, which has any relationship to what we understand by 'faith'. In biblical mythology, the paragon of faith is Abraham. He is the archetype of a believer. The act of believing requires a kind of paradoxical trust, a confidence against evidence. What does Abraham believe? Certainly not that God exists. This he knows for sure. And there's nothing contra-evidential about that. What he believes is the truth of God's promise to make him the ancestor of a great nation, and, as a first step in this direction, that Sara will give birth to a son. We must distinguish between existential or ontological truths and promissory or futurological truths. Ontological truths – 'to be or not to be' – are the object of knowledge. Promissory truths are the object of faith. The promise made to Abraham is indeed contra-evidential, since Sara is 80 or 90 years of age. It is verging on the absurd, so as to make one laugh. Therefore the son will be called Yitzhak, meaning 'he laughed'. Sara laughed, but Abraham believed. And he continued to do so even when God ordered what Abraham could not but understand as the sacrifice of this same son. Faith, therefore, is linked to the future, to a promise to be fulfilled in the future. The truth of faith does not concern the past – for instance, the life of Jesus – but the future. What could this mean in the case of Jesus? Did he make any promises? Is there something contra-evidential which excludes knowing and requires believing? Of course. But in Jesus's case, two answers are possible: one of an ontological, and several others of a promissory character, relating to the future. The ontological answer is that Jesus is the Christ, the Messiah, the Son of God. The promissory answer is that the Kingdom of God will come and that we will be delivered of our sins and will be resurrected from death when that kingdom comes. Faith does not relate to the past. The thesis that the historical Jesus is the Jesus of the Gospel does not really hit the point. It refers to the future. And since this promissory character and the futurological turn seems alien to ancient Egyptian religion and to every other so-called pagan religion, there's nothing into which the words *emouna* or *pistis* could be translated. The turn towards the future seems to be the hallmark of biblical monotheism and its elaborations into Christianity and Islam. There is an eschatological and at times even apocalyptical element inherent in the concept of faith and the related concept of truth. And this may be also the reason why this concept is potentially dangerous, destructive and violent.

Sergio Romano:

I would like to introduce Dr Brian Hebblethwaite, who has taught philosophy of religion at the Faculty of Divinity at Cambridge University, and is a Life Fellow of Queen's College there after having served as Canon Theologian of Leicester Cathedral.

Brian Hebblethwaite:

I should like to begin by making some remarks from the perspective of our theology and religion group about all our discussions so far in these two days. In the first place, I think that most of us have agreed that our very different truth claims in our very varied domains are aimed at truth, at stating and acting on how things are in our very different areas of concern while acknowledging our very partial and limited success in getting at the truth. But the point I want to raise is, that all these areas of concern are interconnected. The physical universe, with which cosmologists, physicists and mathematicians are concerned, has evolved conditions hospitable to and productive of the life forms with which the biologists are concerned, and biological evolution has come up, here on earth, with consciousness, mind, personality, freedom, creativity and culture, with all of which the humanities are concerned. Among these products of cosmic and biological evolution are all the art forms – literature, painting, sculpture, architecture and music, which even Nietzsche, following Schopenhauer, called an *opus metaphysicum*. Science, philosophy and religion, including religious experience, are themselves aspects of what has eventually emerged from cosmic and biological evolution. And we in philosophical theology find ourselves pressing the question: "What does it tell us about reality, about the universe, that it has it in it to come up with all this?" The religions, as you know, have views about that matter.

Now, of course we shall be told that in theology and religion, we are hopelessly divided amongst ourselves, we Christians, Jews and Muslims in the traditions of Semitic origin, to say nothing of the Eastern and Far Eastern religions. Consensus is even more difficult in religion than in the subject areas already covered in this symposium – though we have seen many examples of disagreement between the philosophers, between the mathematicians and the philosophers, between the scientists and cosmologists, and between them and the philosophers too. And the historians don't appear to agree with each other either. We've also heard about our ignorance – ignorance about dark matter or energy. Even Jean-Pierre Changeux began by saying we don't know much about how the brain has come up with consciousness, though he went on to tell us an awful lot about it, and I must say I think I learned more in those forty minutes than I've learned in any forty minutes previously in my life. We also heard about our still being in Plato's cave, and about the inexpressibility of truth. So it should not surprise you that theologians talk about the incomprehensibility of God, or that people of faith admit with St Paul that "now, we see in a glass darkly". We people of religion are accused of being dogmatic. That's not how I understand the exploratory and self-critical character of philosophical theology, and to my mind the most dogmatic views come from those who would restrict attention to the measurable, like Kelvin did, to the material or to the contingent. And the most dogmatic of all are those who accuse all religion and theology of being mumbo-jumbo.

Simon Blackburn mentioned Pilate's question, "What is truth?". Taken in isolation, that question certainly deserves Blackburn's response. It depends on what you're talking about. But in its context, it was an ironic, sceptical politician's refusal to take seriously the claim of Jesus to have come to bear witness to the truth, presumably the truth about God and God's purpose for mankind. And that leaves me to echo what Professor Assmann has just said about truth and religion not being simply about how things are or were, but about how they were meant to be and how they will be. The eschatological dimension is central, certainly, to the religions of Semitic origin. I can't resist just saying – this is very naughty – that I look forward to meeting up with Simon Blackburn in the afterlife and telling him "I told you so", but I think that just goes to show I'm not very far on the path of virtue, and no doubt we'll have better things to think and talk about.

Anyway, I think it's quite implausible to dismiss all theology as mumbo-jumbo. There is a lot of mumbo-jumbo about in religion and theology, but that's true of philosophy too. But it is certainly not true of the great medieval theologians like Maimonides or Thomas Aquinas, so I'm going to defend St Thomas about the sun and the earth or the time of ensoulment. And it's certainly not true, just to mention one or two names, of Joseph Butler in the eighteenth century, of John Henry Newman in the nineteenth century, or of William Temple, Karl Rahner, Alvin Plantinga and William Alston, to name but a few, in the twentieth century. These thinkers speak to us out of living traditions of faith, but what I want to stress is that these traditions, including their revelation claims, are not necessarily just authoritarian systems. They are all, including their revelation claims, open to rational scrutiny and defence. And there are traditions which, among other things, offer wide-ranging worldviews which, to my mind, make better sense of all aspects of our extraordinary universe than those that restrict themselves to certain aspects of the universe, the more easily measurable ones, for instance. I think it's a mistake to think of the religions as just authoritarian. After all, when it was said that Jesus spoke with authority and not as the scribes, that is because of what he said and how he said it, and how it came home to them. And that's true, I think, of the Bible. Samuel Taylor Coleridge, in his writings on the Bible, said that the Bible speaks to us when it *finds* me, he said; when it speaks to the heart and the mind; not quoting some isolated text or whatever, but when the central message finds me and grabs me. That's the sense of authority that I think we need in religion.

Just the briefest of brief comments on my colleagues' presentations in this group. The Cardinal, of course, speaks out of a long, wide-ranging tradition of Christian faith, a tradition that includes the major figures of St Thomas, Newman and Rahner, and he takes seriously, as they did, the revelation claims of Christianity. But he also spoke, as the present Pope and his predecessor have done, of faith and reason. And I repeat, the openness of faith traditions, including their revelation claims, to rational scrutiny is something we should all be prepared to consider, both from within and from without the faith communities, in dialogue. Now I'm not well-placed to comment on the history or theology of Islam, though I have read and profited from the comparative theology of writers such as Kenneth Cracknell and Keith Ward, who've shown something of what can come out of interfaith dialogue between Christians and Muslims. Professor Arkoun mentioned this morning the question of Sharia law, and he will know that the Archbishop of Canterbury in my country got into some hot water recently by suggesting to a group of lawyers that English law should make room for elements of Sharia law. That's another example of interfaith dialogue. And I think there's a lot going for looking for commonalities or complementarities or possibilities of working and living together from the different faith communities in our multi-faith societies of today.

As for the question of violence, I'm very glad to hear what the Cardinal and Professor Arkoun said about that. I think of Archbishop Desmond Tutu and his Truth and Reconciliation Committee in South Africa as a wonderful example of how the question of truth in religion can operate – the very reverse of violence.

To Geza Vermes and his fascinating work on the historical Jesus, let me make just one or two comments about faith and history. Historical judgments on religious matters are of course heavily conditioned by the historian's own religious or non-religious perspectives. I can't resist telling you a story of the nineteenth and twentieth-century historian, T.R. Glover. Two fellows of his Cambridge college met at the college gate. One said to the other "Have you seen Glover's autobiography?" The other replied "I didn't know Glover had written an autobiography. What's it called?" "It's called *The Jesus of History*", the other replied. Attempts to recover the historical Jesus notoriously reflect the writer's own position. I much admire E.P. Sanders's book on the historical Jesus. His perspective is that of a not unsympathetic non-believer. I much admire Geza Vermes's writings on this theme. His perspective is that of a Jewish historian, and it probably gets closer to the facts than Sanders's work. But Christian historians are bound to read the Gospel records differently, coming as they do from participation in the faith community that goes back to Christ crucified and risen. And as I've been stressing – and this is what I want to leave you with, really – the beliefs summarized in the Christian creeds are all subject to rational scrutiny; scrutiny that must take into account what purports to be background knowledge.

Take the question of the resurrection of Jesus, about which Geza Vermes has recently written a book. Obviously, a purely secular historian is not going to give credibility to the resurrection narratives in the Gospels. I'm not going to go all the way with the Oxford philosopher of religion, Richard Swinburne, who in his book on the resurrection of Jesus Christ claims that, given belief in God, and given belief in the incarnation, the resurrection of Jesus has a probability according to Bayes's Theorem of 0.97 – I think that was his figure. Swinburne may be wrong about this, but even his writings on the philosophy of religion are not mumbo-jumbo.

Sergio Romano:

We have custodians of temples and custodians of time. I am the custodian of time. Can you please raise your hand if you want to ask a question.

Comment (Simon Blackburn):

Apropos of meeting Brian Hebblethwaite in the afterlife, I'd just like to comment that there's a nice remark by C.D. Broad, an English philosopher, in one of his books. He is talking about life after death, and he finishes by saying "On this matter, as on so many others, we can only wait and see; or alternatively, which is no less likely, wait and not see".

Question:

Professor Vermes, concerning George Lamsa's English translations from the Aramaic of the Peshitta which he did about 35 years ago: yet we hardly hear of that. Do you know anything about its influence?

Geza Vermes:

I don't think that much has been said about that book, which is naturally totally subjective. Any study of any Aramaic translation of the Gospels will have to decide, first of all, which kind of Aramaic dialect applies. And on this there is no agreement among experts. I myself think that the only kind of Aramaic dialect that would be suitable for any kind of interpretation would be the Galilean dialect of Aramaic, of which we have some sources belonging to a period which is several centuries after the time of Jesus. So it will become very, very subjective, and very, very relative. I don't think anybody will get very far along those lines.

Question:

Professor Vermes, I know of a book explaining Jesus in relation to the Zealots. I don't know if that was the theory of Samuel Brandon. How does it fit with your findings? And also I would like to ask Professor Arkoun a question about violence. I found information that at least one intellectual in the Nazi period, I don't recall his name, decided that Jesus was a Jew but not psychologically so, in order to justify the repression of the Jews. This points to the repression of the Jewish mentality by the Nazi regime. Maybe you have something to say about the difference of this repression, which is purely based on race, and the repression of a mentality that took place then and might take place in our time too.

Geza Vermes:

Of course Jesus and the Zealots is a question that has been debated after Brandon and Hyam Maccoby and some other recent scholars. I myself believe that it's completely mistaken and misunderstood. I don't believe that Jesus was animated by political considerations. And with the exception of one or two sayings that have crept into the Gospels, and with the exception of the possible political significance of the fact that one of the apostles was referred to as 'the Zealot', we have really nothing to go on. It would seem that Jesus was essentially non-political, and that his concept of the kingdom of God was not one that would be achieved by the force of arms.

Mohammed Arkoun:

There are lots of texts which have been published to legitimize these unacceptable policies. Let me just say that the idea that Jesus is not a true Jew is one of those collective inventions, which evolves in the social imagination and might find supporters. You can find any such ideas in all societies as there are always racist people who tend to cast other people out. So we need to be aware of all this, but it's quite a different thing to try and explain these ideas.

Question:

Let me just ask Professor Assmann: the fact that you don't hold the distinction between faith and hope for truth is valid, is this absence of distinction misleading? You seem to confuse faith with hope, whereas they are two well diversified dimensions.

Jan Assmann:

Of course, hope and faith are two different phenomena, as St Paul already distinguished them, and both relate to the future, to a promise, to fulfillment to be trusted in, and trust relates to the authority in which the trust is posed. We spoke of this authority. The German word *Verheissung* is much stronger than 'promise' in English. But promise is a performative speech act, and performative speech acts require the authority of the speaker. And so the highest authority of course is God, or the Son of God, and speaking with authority means being in a position to make a very great promise, which of course is the object both of faith and of hope. So I cannot see the difference between trust and faith. It relates to the authority and to the performative reality of the promise.

Georges Cardinal Cottier:

Let me just refer to the words of Dr Hebblethwaite about the interpretation of the holy texts. For believers and non-believers, there are some biases that we need to consider. An absolutely unbiased and scientific positivist look is not something you can actually have. There are very interesting points in the words of Professor Vermes, but there are several questions raised here. If Christ has been embodied, then He has taken on a new language and new shape. Let me address another question: what is the relationship among cultures when we are facing the problem of tradition, for instance? There are very interesting points of view about the truth in Islam and about the transfer of knowledge in the Arab world. The holy texts should be framed within the historical context as well as in the context of the communities. Dr Hebblethwaite has really highlighted a very important topic there.

Geza Vermes:

This is a half-serious question to Cardinal Cottier. Am I right in remembering that, in your paper, the word 'Jesus' has never appeared?

Georges Cardinal Cottier:

I have just mentioned Jesus Christ twice. But in my paper I started with commenting a text which was actually not about Him, as I started from the Gospel of John.

Sergio Romano:

I think we ought to thank the participants at the roundtable and the main speakers, because we have been given a lot of food for thought and we ought to be very grateful to you all.

Session VI: Conclusion

17 May 4:00 p.m.

Chair:
Werner Stauffacher

Synthesis:
'Truth': Common Themes?
Lord Krebs and Simon Blackburn

Comments:
Mike Gray and Alessio Marconi

Closing remarks:
Bruno Bottai and Achille Casanova

Werner Stauffacher:

You have probably realized that I am not Professor Decleva, who was meant to chair this final session. He was held back in Milan at the last minute due to very urgent and important official obligations, and sends his heartfelt apologies.

I am very happy to see the amphitheatre still full at this late hour of the second day of a demanding symposium and should like to thank all the speakers and participants who are still here for having stayed up to the end. This is true, in particular, for the young scientists and students attending the Symposium as invitees of the Italian and Swiss Academies and whose active interest and participation in the discussions throughout the past two days were exemplary and a pleasure to witness. Thank you very much for this.

We are now concluding this conference with a summary given by John Krebs and Simon Blackburn. I have been asked why we choose these two speakers and why the concluding remarks should be in English. The answer is quite simple. Lord Krebs has suggested the title of this conference, i.e. 'Truth', about two years ago after having read Simon Blackburn's book with the same title. Thus, Simon Blackburn and John Krebs were at the very origin of this symposium. We therefore found it appropriate to hear their impressions after two days of enlightening talks

Balzan (ed.), *Truth in Science, the Humanities and Religion: Balzan Symposium 2008*,
DOI 10.1007/978-1-4020-9896-3_7, © Springer Science+Business Media B.V. 2010

and very lively debates on this elusive topic. And it is only logical that they should present it in their mother-tongue.

The title they have chosen for their synthesis is:

'Truth': Common Themes?

Lord Krebs:

I had been going to provide a grand and comprehensive overview under the title 'The Truth about Truth', but when I heard Keith Thomas this morning telling us that it was impossible to capture what had been said in these last two days, I gave up. It was impossible, partly because of the unreliability of memory, and partly because many of the important thoughts will have been private thoughts and conclusions that each of us have had. So instead, what I want to do in these few minutes, is try and look back over the two days and reflect on the fact that we've been very fortunate to hear excellent presentations across an unusually wide range of disciplines and perspectives. The question that was in my mind is this: as we've gone through the two days, has there been convergence or divergence of thought about what is meant by truth from these different perspectives?

So I'm going to talk first of all about *kinds of truth.* Simon Blackburn, in his keynote lecture, drew a distinction between absolute truth and relative truth – the idea of cultural relativism. And in setting up the argument, he drew a rough equation between absolute truth being associated with the natural sciences – natural scientists are discovering things that are absolutely true – and relative truth, being more characteristic of the humanities. But as the two days have proceeded, I think we've seen various ways in which that distinction has become slightly more textured. And particularly, I think, in this morning's session Quentin Skinner and Dominique Schnapper both modified the idea of *conceptual* relativism, in which there is absolutely no basis for distinguishing between different versions of opinions about the truth, and *contextual* relativism. In Quentin's talk he said the peasants of Languedoc may have had a rational basis for their beliefs on the basis of what they understood at that time. Dominique Schnapper summed this up in a very pithy sentence, "Don't judge the past by the present". But, also this morning, we heard from Lord May, echoing what Bengt Gustafsson had told us yesterday, that the notion of natural science being about a set of established facts you read in a textbook is completely wrong, because much of scientific knowledge is highly provisional, is evolving, is incomplete. Natural science is a way of knowing, rather than a set of established facts. So just as Dominique Schnapper said of history, "Don't judge the peasants of Languedoc of the past by the present", you could say of scientists, "Don't judge past scientists by present standards". Don't criticize Darwin because he knew nothing about the laws of genetics, even though they'd already been discovered by Gregor Mendel. So later in his life he couldn't get natural selection as a mechanism of evolution to work, because he thought inheritance was blending, and that we'd all end up identical, while Mendel of course discovered that inheritance is particulate. So there are some convergences there, and in one sense, science is contextually relative.

Moving on to my second brief point about *sources of truth*. Lord May introduced his talk by referring to, in his opinion, three kinds of truth, according to their sources. Natural scientific truth which is drawn from evidence – and we heard from Bengt Gustafsson what evidence implies: observation, accurate measurement, quantification, replication and challenge – and the sister of the natural sciences, mathematics, where truths are derived by reasoning. And those were contrasted with truth derived in the case of religion or beliefs or values which come from texts or authorities, or revealed truths in Geza Vermes's term this afternoon, the 'gospel truth'. Now, having heard this afternoon's session, one could say that that distinction is still firmly there; that much of this afternoon's session drew on the authority of ancient texts to establish the truth. However, I think Keith Thomas made an important point this morning – although he didn't quite put it in these terms: today, scientists are the new priests. Most of you, and as Keith Thomas confessed himself, have not actually measured that the moon is a quarter of a million miles away. You accept that because a priest-like scientist has told you, and you have trust in the authority of the scientist. So the question that Keith posed about this issue of belief, or truth based on revealed truth, texts and authority is: "Whose authority do you trust?" And as he said, when there are competing authorities – perhaps the authority of the Bible or religious texts on the one hand and the authority of the scientist on the other – it is an interesting question for society as to which authority one follows.

My third comment, very briefly, is about the *location of truth*: is truth out there, or is truth in here? We natural scientists tend to accept and believe that we're discovering truths about the natural world that exist independently of us. However, as Jean-Pierre Changeux pointed out so beautifully in his talk, the brain does not act like a video camera simply filming the outside world in a passive way; the brain is an interpreter, a modeller, a creator of possible worlds. And in a sense that also brought in the question of imagination, because what Jean-Pierre Changeux said is that the brain, with its capacity for consciousness, is capable of imagination. I don't have to do the experiment of pouring this glass of water over Simon's head because I can imagine what would happen if I did – he'd stand up and punch me on the nose. So my imagination gave me the reality to put the glass back down. It gave me the truth. So, again, the distinction between truth out there and truth in here is not an absolute one, because our brains act as interpreters and predictors and modellers of the world around us.

My final point is about the *universality of truth*, and Professor Assmann, in his intervention this afternoon, made an interesting and important point which I don't think had been brought up before, and it's this: if, as a natural scientist you establish some fact about the world that fact is true wherever you measure it. The atomic weight of sodium is the same whether it's measured in Chile, in China or in Chad. There isn't a Jewish version of the atomic weight of sodium, there isn't a Christian version, there isn't a Buddhist version. It is a universal fact, a universal truth. As Professor Assmann pointed out, many religious truths don't have this property. They are particular. You have a different version of the truth if you're a Christian, and even within Christianity, if you're a Catholic or a Protestant you have a different version of the truth. So there is something about the universality of truth that distinguishes truths derived from the natural sciences and truths associated with religion.

But what about, and this is my final point, *truths associated with moral beliefs and values*? And here, I think, is a whole area of research which has emerged in the last decade or so which we haven't touched on today and which is extremely interesting. I'll just refer to one study, carried out by Marc Hauser at Harvard together with Peter Singer, in which they created on the web some moral dilemmas: moral choices that you and I could reflect on and come to a view about what is the ethically correct thing to do. They're about balancing harm and fairness, and I won't go into the details. But the crucial thing is this: that they've tested these moral dilemmas on more than 200,000 people from 120 different countries, including remote pre-civilization tribes in the Amazon jungle and the forests of Papua New Guinea, and to a first approximation, the moral dilemmas are answered in the same way by everybody, regardless of cultural background, regardless of whether they're modern civilizations or primitive civilizations, regardless of whether they're religious believers or atheists. So perhaps there are some moral universals, and in order to understand those, we need to look to Darwinian evolution. Those are my conclusions.

Simon Blackburn:

I'm not quite sure how to sum up. So much has been said that has sparked thoughts, but in the way of these things those thoughts are fleeting and evanescent and I probably misremember them. So I'll just pick a few things at random that have interested me.

One was said very late on by Brian Hebblethwaite when he reported Richard Swinburne's Bayesian calculation of the probability of 0.97 of the Resurrection, provided we give the theist the premise that God exists and that Jesus was his son. That strikes me as rather strange because I should have thought that the Bayesian probability would be very low, because we don't know very much about God. But one thing we do know is that most of his creation dies, and very, very few of them seem to survive death, as far as the empirical evidence goes. So I should have thought that the actual prior probability of success in that direction would be vanishingly small. However, I haven't seen Swinburne's calculation, so I can't go into that in more detail.

I'd like to pick up something from this morning, which I think also raises questions which were in the air this afternoon. This was about the issue of faith and rationality. I think Quentin Skinner did a wonderful job of presenting reasons why the historian might find it necessary to concentrate on rationality rather than on truth, and I think he was absolutely right and I thoroughly agree with everything he said. Of course, individual truths go on being topics because we're interested in particular propositions. But what about the nature of truth? As Quentin Skinner said, historians can do without a theory of it, and the surprising thing is that, at least over large tracts of our subject, philosophers can do without a theory of it as well, although we are of course interested in rationality and we're very interested in evidence. So perhaps next time, if in five years time an equivalent conference to this is mooted, the topic might be method, or evidence, which I think is at least equally important if not more so.

I think the interest of rationality for me is very much tied up with the interest of putting people in a kind of situation, seeing them as doing a kind of thing which is truth-conducive. So I think you see people as rational if you see them as – however ineptly or unluckily in some instances – attempting to put themselves into a class of people who've got more chance of being right than being wrong about whatever it is. So Quentin Skinner's Languedoc peasants, insofar as they were listening to the attested authorities of their time and taking their beliefs from them, were doing something rational; they were doing something that we all have to do, actually, because we can only succeed in putting ourselves in the class of people with a high truth ratio if we perform that way. Now, of course, human understanding is apt to become very baffled. Our understanding of each other becomes very baffled when we can't see a group, or a culture, or a subset of people having ever tried to do that or having even come anywhere near doing it. So, for example, those people in San Francisco who decided to commit suicide because they felt that thereby they could join the Hale-Bopp comet which was coming to pick up dead computer operators. One simply cannot get on all fours with that belief – it doesn't seem to have authority, it's got no evidence, it's got absolutely nothing going for it. It seems to me that pure faith, unfortunately, is in that same bracket. If somebody 'just believes' something, and all that can be said is: "He just believes it", then the chances of him being right and the chances of it being true are very low, no better than random. Depending on the prior probability of what they believe, it may be very low indeed. And consequently I think that these sort of random bursts of confidence not only don't deserve respect, but in a sense they baffle understanding. It's very difficult to see what's gone on in somebody's head when he is, as it were, deliberately turning his face away from the procedures which are adapted to winnowing out truth from falsity. Fortunately, I think, over huge tracts of life, very few people are like that – the Hale-Bopp hopefuls were very unusual. I'm sure Quentin Skinner is right that when we attempt the disciplines of human interpretation – the hermeneutic disciplines – we have to proceed under the supposition that people are not like that, and that there can be something to be said for whatever position they found themselves in. And I think that's an important, as it were, charitable motto, both as we try to understand other groups in the world today and, presumably, as we try to understand historical agents. A particularly important rider on that, which was adumbrated this morning by Keith Thomas and which John Krebs referred to a moment ago, is, of course, the issue of authority and the issue of trust. When do human beings deserve confidence? Well, of course we all know certain signs of when human beings deserve confidence. Not just that they themselves are confident – they may have done their stuff, done the right thing; they've been and looked, as we say. There are other cases where it's not so clear why they deserve confidence. Perhaps there's a tradition of according confidence to some text or to some other cultural transmission which has brought down an opinion which, for some reason, we think ought to be trusted. And I think it's one of the glories of science, but equally of history, of the human disciplines – and this is a point of contact – that while we work in traditions, we are also prepared to turn a critical eye upon them. That may be a culturally peculiar thing about the West, I just don't know. But it's a glorious thing wherever it's found. For that reason, I think, there are morals from what we've learned to be drawn for the practice of education. It has long struck me that science education, particularly in the United Kingdom – I don't know about in the rest of Europe – proceeds far too often as if there's just a body of truth which is the teacher's job to pour into the pupil and the pupil's job to receive. And that seems to me absolutely inimical to the spirit of science and to the value of science as a human activity. The child or person may grow up and be little better than a pocket calculator, and about as much use – in other words, without an intelligence at the other end, there's nothing to be proud about. I think science ought to proceed in the light of its own history, and that includes the history of many false steps, many controversies, many cases where great scientists were proved wrong about one thing or another. But in cases where the ship righted itself, or so we think, it should proceed in the light of the philosophy of science, which would include a story about methodology, about what's so good about the experimental method and why it works, and why we have faith that it will continue working. So there's a brief 'please' for the contextualization, if you like, the humanization of science in education. I think this is a message I would certainly like to take away from this weekend, and I hope other people can take it away as well.

That said, I will now shut up. You heard enough from me yesterday. I'd like to thank you all very much for putting up with the language difficulty, if you found it – I echo again Quentin's words that we British have to remember that we are accorded the privilege of other people understanding our language, but we must never take it for granted and must always be grateful for it.

Werner Stauffacher:

Thank you very much, both of you. That was not easy, and it was a wise decision not to attempt to summarize the last two days, to try to put them into a nutshell. It would have been an impossible, not to say superficial attempt. The final verdict on this symposium, on its merits, its shortcomings, will not be in for some time. However, you have both put a finger on potential convergences and remaining divergences and contradictions, and it is on these that we will have to work and concentrate if we want to progress. Your presentations were a worthy conclusion to our debates. I would like to end here, say "thank you", and pass the word to the two Presidents of Balzan's two Boards of the Foundation for their concluding remarks.

Concluding Remarks

Bruno Bottai:

This will not be a speech. I think you have already heard too many. In any case, we must conclude, and this is just a goodbye. Truth is very difficult to attain, and on the way to attaining it there are also a few traps or pitfalls, which may sometimes appear to lead the way to truth, but are not really accurate and may lead – not to truth, but to the contrary of truth, something which is less than truth. This symposium has shown instead that, although it may be very hard to attain truth, if we work hard we can reach it, or we can at least get very to close to it.

The Balzan Foundation has a certain tradition which is in full agreement with this undertaking. Our foundation is a witness to faithfulness and to truth, and the same holds true for its founder. For many years, Eugenio Balzan has been working in Milan as a support to culture at its highest level, regardless of immediate results. As a bi-national foundation, the Balzan Foundation plays this role – and let us not forget this: it also fulfills this function between Italy and Switzerland, and in particular between Milan and Canton Ticino.

I would like to thank the public attending our debates for following us. I also thank our host authorities here at the University and those of the city of Lugano, who have so kindly helped us. Mrs. Giovanna Masoni Brenni, Municipal executive of the city, is now with us, and, through her, I would like to thank all the people of the city of Lugano who have helped us.

Achille Casanova:

Just a few words to associate myself with what has just been said by Ambassador Bottai. Also the Balzan Foundation "Fund" in Zurich found this initiative very interesting and we supported it with enthusiasm. From my point of view a symposium on truth is a courageous, even bold enterprise. It was a real challenge, but I think this challenge was met successfully, and in these two days we were able to learn a lot. You should know that the Balzan Foundation, while honouring established academics for their work, is peculiar in that it also supports young researchers. Indeed, half of the Prize-money of each Balzan Prize must be devoted to research projects involving young academics. In this same line, we thought that it would be appropriate to invite young doctoral students from Italian and Swiss universities to this symposium. It would be nice if one or two of these students would now come forward and tell us very briefly what their impressions were of these two days spent 'in search of truth'.

Mike Gray (University of Zurich, Institute for Hermeneutics and Religious Studies):

Listening to Quentin Skinner today, I was enormously impressed by his generosity towards people of the past who thought and believed things that we find silly today. Listening to Keith Thomas, I was even more impressed at the way he extended his generosity to the present day: to people in places like Lourdes, where people have ideas many of us would not agree with; perhaps even to people in places like Kansas, or even perhaps to all of us Americans, who knows? However, as I reflected on this, it seemed to me we should extend still greater generosity, even to such people as historians, who come up with silly ideas or anachronistic interpretations, with judgments of the past and with reasons why people did the things they did, and do so based on the knowledge we have today.

In *Anna Karenina*, Tolstoy talks about a man named Stiva Oblonsky who never got over-excited at work and never made mistakes or talked nonsense, because he didn't care. Now, judging by the vigorous critical reaction that all of you, speakers, got from us doctoral students yesterday and today, I think you all cared deeply. And judging by how much debate there was, including the occasional misunderstandings, taking into account the various questions we asked and discussed among us, I think we cared too. So, in a spirit of generosity and thankfulness I would like to say thank you to all of you who spoke at this conference, for caring enough to transmit some of your care to us and give us stimulus to care for ourselves, in the future, for the questions that we're struggling with in our various areas. And thank you very much, Balzan Foundation, for giving us the possibility to be here.

Alessio Marconi (Milan State University, Institute of Philosophy):

As a simple philosophy student I cannot hide a certain embarrassment after having met such impressive names who took the floor in this symposium. Without entering too much into details, I will simply give my general impression on the results: it is not simple to look for truth. We have noticed this in the various sessions and through the different points of view developed in this search for truth. Debates have been going on for millennia, perhaps ever since mankind exists. Maybe, the difficulty is brought about – take the example of the discussions about cosmology – by the need to study something which is not easy to compare with known systems from which truth could be deduced. Or perhaps it's truth about something which no longer exists, something which is absent. This is what historians are engaged in: searching for truth in what has survived from a period which no longer exists.

In this sense, I feel obliged to say, these two wonderful days could not give me clear and certain answers, let alone *one* certain answer, as to what truth may be. But they certainly help to better formulate the questions. Also, thanks to the multidisciplinary approach of the discussions, where looking at the same thing from different points of view was the norm – both between different disciplines and within individual disciplines – we may not leave this symposium with one answer, but we may be able to leave with better answers, with better questions, enriched by something which we had not taken into consideration in an adequate way beforehand. The points raised in this symposium will enable us to come up with questions of enriched quality. This allows us to take a further step, no matter how small, but necessary to improve our knowledge and to widen our horizons. Therefore I would like to thank wholeheartedly all those who took part in this symposium, the speakers and all the others who made this symposium possible.

Achille Casanova:

Thank you very much, both of you. I have said a few moments ago that I had the impression that this symposium's bold challenge of speaking about truth had been met successfully, and, in my view, these two students' statements confirm this impression. To me, the last statement was particularly important. We have been able to face the theme of truth by asking the right questions, and asking the right questions, I think, is essential in human development.

It is time for me to thank all of you who took part in this symposium, to thank of course those who animated it, all the personalities who took part in it, to thank those who organized it. I would like to thank the members of the preparatory group of the Foundation's General Prize Committee, who for months worked on the planning of our sessions, finding and choosing the right people in order to balance the session's content from various philosophical, scientific and theological points of view. They did a fine job and deserve your warm applause.

Of course, I want to thank Mrs Suzanne Werder, Secretary General of the Balzan "Prize" Foundation, who with her collaborators has been *the* central organizer, a perfect organizer of this symposium. I would like to thank the University and the City of Lugano. The city, unfortunately, presented itself in a unfamiliar way, because, I can assure you, Lugano is usually a very sunny city in May, and such heavy and continuous rain as we had is extremely rare. Maybe, this constraint posed by the weather has had a positive effect, in the sense that it has not tempted us to leave this room to go visit beautiful Lugano – but I assure you, this was not our intention.

Thank you for coming and for participating in this symposium, and I hope to see you at the next one.

Bibliography

Simon Blackburn *Truth and Ourselves: The Elusive Last Word*

James, William *The Essential Writings* (New York 1971).
Nietzsche, Friedrich Wilhelm *Beyond Good and Evil: Prelude to a Philosophy of the Future,* in: *Basic Writings of Nietzsche*. Transl. by Walter Kaufmann (New York 2000).
Nietzsche, Friedrich Wilhelm *The Twilight of the Gods,* in: *The Portable Nietzsche*. Transl. by Walter Kaufmann (New York 1968).

Marcelo Sánchez Sorondo *discussion panel Session I: Philosophy*

Hegel, Georg Wilhelm Friedrich *Enzyklopädie der philosophischen Wissenschaften im Grundrisse (1830)*. Ed. by Udo Rameil et al. (Hamburg 1992).
Kant, Immanuel *Kritik der praktischen Vernunft*. Ed. by Felix Meiner (Hamburg 1990).
Kant, Immanuel *Kritik der reinen Vernunft*. Ed. by Ingeborg Heidemann (Leipzig 1956).
Kant, Immanuel *Werke*. Ed. by Ernst Cassirer et al. (9 vols.; Berlin 1912–1922).
St Thomas Aquinas, cf. http://www.josephkenny.joyeurs.com/CDtexts/index.htm.
St Thomas Aquinas, *Summa Theologica* (3 vols.; New York 1947).

Yuri Manin *Truth as a Value and as a Duty: Lessons of Mathematics*

Bovens, Luc – Stephan Hartmann *Bayesian Epistemology* (Oxford 2003).
Clements, Richard R. – Roger L. Hughes 'Mathematical Modelling of a Mediaeval Battle: the Battle of Agincourt 1415,' *Mathematics and Computers in Simulation* 64 (2004) 259–269.
Dauben, Joseph Warren *Georg Cantor. His Mathematics and Philosophy of the Infinite* (Princeton 1990).
Davis, Philip J. – Reuben Hersh *The Mathematical Experience* (Boston 1981).
Enzensberger, Hans Magnus *Drawbridge Up. Mathematics – a Cultural Anathema* (Natick 1999).
Lakoff, George 'The Contemporary Theory of Metaphor', in: Andrew Ortony ed., *Metaphor and Thought* (2nd ed.; Cambridge 1993) 202–251.
Mandelbrot, Benoît B. – Richard L. Hudson *The (Mis)behavior of Markets: a Fractal View of Risk, Ruin, and Reward* (London 2004).
Manin, Yuri I. *Mathematics as Metaphor. Selected Essays* (Providence, RI 2007).
Manin, Yuri 'Mathematical Knowledge: Internal, Social and Cultural Aspects', in: Yuri I. Manin, *Mathematics as Metaphor. Selected Essays* (Providence, RI 2007) 3–26.
Peters, John Durham *Courting the Abyss: Free Speech and the Liberal Tradition* (Chicago 2005).
Poovey, Mary 'Can Numbers Ensure Honesty? Unrealistic Expectations and the US Accounting Scandal', *Notices of the American Mathematical Society* 50 (2003) 27–35.
Stewart, Ian *The Problems of Mathematics* (Oxford 1987).

Bengt Gustafsson *Thoughts on Truth in Physics, Astrophysics and Cosmology*

Ferris, Timothy *The Red Limit: the Search for the Edge of the Universe* (New York 1977).
Feynman, Richard P. *The Character of Physical Law* (London 1967).
Pickering, Andrew *Constructing Quarks: a Sociological History of Particle Physics* (Chicago 1984).
Shapin, Steven *A Social History of Truth: Civility and Science in Seventeenth-Century England* (Chicago 1994).
Shapin, Steven – Simon Schaffer *Leviathan and the Air-Pump: Hobbes, Boyle, and the Experimental Life* (Princeton 1985).
Sibum, Heinz Otto 'Reworking the Mechanical Value of Heat: Instruments of Precision and Gestures of Accuracy in Early Victorian England', *Studies in History and Philosophy of Science* 26 (1995) 73–106.
Sibum, Heinz Otto 'The Number of the Century: A History of a Scientific Fact', in: *Wetenschap en samenleving: Erasmusprijs 2005 = Science and society: Erasmus Prize 2005* (Amsterdam 2006) 61–94.
Torretti, Roberto *The Philosophy of Physics* (Cambridge 1999).
Weinberg, Steven 'Conceptual Foundations of the Unified Theory of Weak and Electromagnetic Interactions', in: Stig Lundqvist ed., *Nobel Lectures, Physics 1971-1980* (Singapore 1992) 543–559.
Wigner, Eugene P. 'The Unreasonable Effectiveness of Mathematics in the Natural Sciences,' *Communications in Pure and Applied Mathematics* 13 (1960) 1–14.
Ziman, John Michael *The Force of Knowledge: the Scientific Dimension of Society* (Cambridge 1976).

Jean-Pierre Changeux *Is the Brain the Organ of Truth?*

Changeux, Jean-Pierre The *Physiology of Truth: Neuroscience and Human Knowledge* (Cambridge, MA 2004).
Changeux, Jean-Pierre *Du vrai, du beau, du bien : une nouvelle approche neuronale* (Paris 2008).
Changeux, Jean-Pierre – Paul Ricœur *What Makes Us Think?: a Neuroscientist and a Philosopher Argue about Ethics, Human Nature, and the Brain* (Princeton 2000).

Quentin Skinner *Truth and Explanation in History*

Clark, Stuart 'Inversion, Misrule and the Meaning of Witchcraft', *Past & Present* 87 (1980) 98–127.
Cohn, Norman *Europe's Inner Demons* (London 1976).
Davidson, Donald 'On the Very Idea of a Conceptual Scheme', in: Donald Davidson, *Inquiries into Truth and Interpretation* (Oxford 1984) 183–198.
Hollis, Martin 'The Social Destruction of Reality', in: Martin Hollis – Steven Lukes eds., *Rationality and Relativism* (London 1982) 67–86.
Le Roy Ladurie, Emmanuel *The Peasants of Languedoc*. Transl. by John Day (London 1974).
Macdonald, Graham – Philip Pettit *Semantics and Social Science* (London 1981).
Veyne, Paul *Did the Greeks Believe in their Myths*. Transl. by Paula Wissing (Chicago 1988).

Dominique Schnapper *Cultural Relativism: Meaning and Limits*

Abou, Selim *Retour au Parana. Chronique de deux villages guaranis* (Paris 1993).
Aron, Raymond *La révolution introuvable. Réflexions sur les événements de mai* (Paris 1968).
Aron, Raymond *Introduction à la philosophie de l'histoire. Essai sur les limites de l'objectivité historique*. Ed. by Sylvie Mesure (Paris 1986).

Baudelot, Christian – Roger Establet *L'école capitaliste en France* (Paris 1971).
Bourdieu, Pierre – Jean-Claude Passeron *Les héritiers. Les étudiants et la culture* (Paris 1964).
Bourdieu, Pierre – Jean-Claude Passeron, *La reproduction. Eléments pour une théorie du système d'enseignement* (Paris 1970).
Dubet, François – Danilo Martucelli *A l'école. Sociologie de l'expérience scolaire* (Paris 1996).
Hamon, Léo *Les juges et la loi. Naissance et rôle d'un contre-pouvoir: Le Conseil Constitutionnel* (Paris 1987).
Montesquieu, *De l'esprit des lois,* in: Montesquieu, *Oeuvres complètes* II. Ed. by Roger Caillois (Paris 1986).
Prost, Antoine *L'enseignement s'est-il démocratisé? Les élèves des lycées et collèges de l'agglomeration d'Orléans de 1945 `a 1990* (Paris 1986).
Semelin, Jacques *Purifier et détruire. Usages politiques des massacres et génocides* (Paris 2005).
Van Zanten, Agnès *L'école de la périphérie. Scolarité et ségrégation en banlieue* (Paris 2001).
Weber, Max *Essais sur la théorie de la science* (Paris 1992).

Georges Cardinal Cottier OP *Faith and Truth*

Cottier, Georges *Mémoire et repentance. Pourquoi l'Eglise demande pardon* (Paris 1998).
Dignitatis humanae. Declaration on religious freedom (1965) cf. http://www.vatican.va/archive/hist_councils/ii_vatican_council/documents/vat-ii_decl_19651207_dignitatis-humanae_en.html, n.1.
Pascal, Blaise *Pensées et opuscules.* Ed. by Léon Brunschvicg (Paris 1978) 587.
Ratzinger, Joseph (Pope Benedict XVI), *Jesus of Nazareth* (London 2007).

Mohammed Arkoun *Violence, the Sacred, and the Regimes of Truth*

Arkoun, Mohammed *Essais sur la pensée islamique* (Paris 1973).
Arkoun, Mohammed *Lectures du Coran* (Paris 1982).
Arkoun, Mohammed *Pour une Critique de la Raison Islamique* (Paris 1984).
Arkoun, Mohammed *The Unthought in Contemporary Islamic Thought* (London 2002).
Arkoun, Mohammed *De Manhattan à Bagdad. Au-delà du Bien du Mal* (Paris 2003).
Arkoun, Mohammed *Humanisme et islam. Combats et propositions* (3rd ed.; Paris 2006).
Arkoun, Mohammed *ABC de l'Islam: pour sortir des clôtures dogmatiques* (Paris 2007).
Changeux, Jean-Pierre *L'homme neuronal* (Paris 1983).
Changeux, Jean-Pierre *Du vrai, du beau, du bien: une nouvelle approche neuronale* (Paris 2008).
Vattimo, Gianni *Après la chrétientè. Pour une christianisme non religieux* (Paris 2004).
Veyne, Paul *Foucault, sa pensée, sa personne* (Paris 2008).

Geza Vermes *The Truth about the Historical Jesus*

Allison, Dale C. *Jesus of Nazareth: Millenarian Prophet* (Minneapolis 1998).
Bultmann, Rudolf *Jesus and the Word* (1st ed. 1926; London 1962).
Bultmann, Rudolf *History of the Synoptic Tradition* (1st ed. 1921; Oxford 1963).
Casey, Maurice *From Jewish Prophet to Gentile God: The Origins and Development of New Testament Christology* (Cambridge 1991).
Crossan, John Dominic *The Historical Jesus: The Life of a Mediterranean Jewish Peasant* (New York 1991).
Dunn, James D.G. *Jesus Remembered* (Grand Rapids 2003).
Fredriksen, Paula *Jesus of Nazareth, King of the Jews: a Jewish Life and the Emergence of Christianity* (New York 1999).
Harvey, Anthony Ernest *Jesus and the Constraints of History* (London 1982).

Meier, John P. *A Marginal Jew: Rethinking the Historical Jesus* (4 vols.; New York and New Haven 1991–2009).
Perrin, Norman *Rediscovering the Teaching of Jesus* (London 1967).
Ratzinger, Joseph Pope Benedict XVI, *Jesus of Nazareth* (London 2007).
Sanders, E.P. *Jesus and Judaism* (London 1985).
Sanders, E.P. *The Historical Figure of Jesus* (London 1993).
Schweitzer, Albert *The Quest of the Historical Jesus* (1st ed. 1906; London 2000).
Vermes, Geza *Jesus the Jew: A Historian's Reading of the Gospels* (1st ed. 1973; London 2001).
Vermes, Geza *The Changing Faces of Jesus* (London 2000).
Vermes, Geza *The Authentic Gospel of Jesus* (London 2003).

Chairmen, Speakers, Invited Discussants

Mohammed Arkoun (France) – Emeritus Professor, La Sorbonne, Paris, Senior Research Fellow and member of the Board of Governors, The Institute of Ismaili Studies (IIS), London

Jan Assmann (Germany) – Professor of Egyptology, University of Heidelberg

Marco Baggiolini (Switzerland) – Director *a.i.* of the Swiss National Supercomputing Centre (CSCS), former President of the University of Lugano, M.D., biochemist

Mauro Baranzini (Switzerland) – Professor of Political Economy, Dean of the Faculty of Economics, University of Lugano

Simon Blackburn, FBA (UK) – Professor of Philosophy, University of Cambridge

Bruno Bottai (Italy) – *Chairman of the Board of the International Balzan "Prize" Foundation*, Ambassador, President of the Società Dante Alighieri

Achille Casanova (Switzerland) – *Chairman of the Board of the International Balzan "Fund" Foundation*, former Vice-Chancellor of the Swiss Confederation

Jean-Pierre Changeux (France) – *2001 Balzan Prize for Cognitive Neurosciences,* Professor Emeritus, Collège de France, former Director of the Research Unit on Receptor and Cognition, Pasteur Institute

Georges Cardinal Cottier, OP (Switzerland) – Pro-Theologian Emeritus of the Pontifical Household, Vatican City

Paolo de Bernardis (Italy) – *2006 Balzan Prize for Observational Astronomy and Astrophysics* (with Andrew Lange), Professor of Astrophysics and Observational Cosmology, University of Rome 'La Sapienza'

Pascal Engel (Switzerland) – Professor of Modern and Contemporary Philosophy, University of Geneva

Giorgio Giudici (Switzerland) – Mayor of Lugano, architect

Mike Gray (USA) – University of Zurich, Institute for Hermeneutics and Religious Studies

Bengt Gustafsson (Sweden) – Professor of Theoretical Astrophysics, Uppsala University

Kai Hauser (Germany) – ICREA Research Professor, University of Barcelona, mathematician and philosopher

Brian Hebblethwaite (UK) – Life Fellow of Queens' College, Cambridge, philosopher of religion

Lord Krebs of Wytham, FRS (UK) – *Member of the General Prize Committee of the International Balzan "Prize" Foundation,* Principal of Jesus College, Oxford, former Royal Society Research Professor, University of Oxford

Nicole Le Douarin (France) – *Member of the General Prize Committee of the International Balzan "Prize" Foundation,* Member of the Institut de France, Honorary Permanent Secretary of the Académie des Sciences, Honorary Professor at the Collège de France

Per Olof Lindblad (Sweden) – *Vice-Chairman of the General Prize Committee of the International Balzan "Prize" Foundation*, Professor Emeritus of Astronomy at Stockholm University, Member of the Royal Swedish Academy of Sciences

Yuri Manin (Russia/Germany) – Professor of Mathematics, Max-Planck-Institute for Mathematics, Trustee Chair and Professor of Mathematics, Northwestern University

Alessio Marconi (Italy) – Milan State University, Institute of Philosophy

Piero Martinoli (Switzerland) – President of the University of Lugano, Professor Emeritus of the Université de Neuchâtel, physicist

Lord May of Oxford, OM AC FRS (Australia/UK) – *1998 Balzan Prize for Biodiversity,* Professor in the Department of Zoology, Oxford University, and at Imperial College, London, Fellow of Merton College, Oxford, former President of the Royal Society, London

M.E.H. Nicolette Mout (Netherlands) – *Vice-Chairwoman of the General Prize Committee of the International Balzan "Prize" Foundation,* Professor of Modern History and Professor of Central European Studies at the University of Leiden, Member of the Royal Netherlands Academy of Arts and Sciences, Foreign Member of the Österreichische Akademie der Wissenschaften

Sergio Romano (Italy) – *Chairman of the General Prize Committee of the International Balzan "Prize" Foundation*, Ambassador and historian

Marcelo Sánchez Sorondo (Vatican City) – Bishop, Professor of History of Philosophy at LUMSA University, Rome, Chancellor of the Pontifical Academy of Sciences

Dominique Schnapper (France) – *2002 Balzan Prize for Sociology*, Research Director at the Ecole des Hautes Etudes en Sciences Sociales (EHESS), Paris, member of the French Conseil Constitutionnel, sociologist

Emanuele Severino (Italy) – Professor Emeritus of Theoretical Philosophy, University of Venice Ca' Foscari, Professor of Fundamental Ontology at the Università Vita-Salute San Raffaele, Milan, Corresponding Member of the Accademia Nazionale dei Lincei, Rome

Quentin Skinner, FBA (UK) – *2006 Balzan Prize for Political Thought: History and Theory*, Barber Beaumont Professor of the Humanities, Queen Mary, University of London, Fellow of Christ's College Cambridge, Foreign Fellow of the Accademia Nazionale dei Lincei, Rome

Werner Stauffacher (Switzerland) – *Vice-Chairman of the General Prize Committee of the International Balzan "Prize" Foundation*, Professor Emeritus of Internal Medicine at the University of Basle, former President of the Swiss Academy of Medical Sciences

Karlheinz Stierle (Germany) – *Member of the General Prize Committee of the International Balzan "Prize" Foundation*, Professor Emeritus of Romance Literatures at the University of Constance, Member of the Heidelberger Akademie der Wissenschaften, Corresponding Fellow of the Académie des Sciences morales et politiques, Paris

Sir Keith Thomas, FBA (UK) – *Member of the General Prize Committee of the International Balzan "Prize" Foundation,* Fellow of All Souls College, Oxford, former President of the British Academy, London, and of Corpus Christi College, Oxford

Salvatore Veca (Italy) – *Member of the General Prize Committee of the International Balzan "Prize" Foundation,* Professor of Political Philosophy and Vice-Chairman of the Institute for Advanced Study of Pavia (IUSS)

Geza Vermes, FBA (UK) – Professor Emeritus of Jewish Studies and Emeritus Fellow of Wolfson College, University of Oxford, Director of the Forum for Qumran Research, Oxford Centre for Hebrew and Jewish Studies

Appendix

The International Balzan Foundation

The International Balzan Foundation was established in Lugano in 1956 thanks to the generosity of Lina Balzan who had come into a considerable inheritance on the death of her father, Eugenio. She destined this wealth to honour her father's memory.

Eugenio Francesco Balzan was born in Badia Polesine, near Rovigo (Northern Italy), on 20 April 1874 into a family of landowners. He spent almost his entire working life at Milan's leading daily paper *Corriere della Sera*. After joining the paper in 1897, he worked his way through, from editorial assistant, to editor-in-chief and then special correspondent. In 1903 editor Luigi Albertini made him managing director of the paper's publishing company; he then became a partner and shareholder in the company. He was not only a clever and skilful manager but also a leading personality in Milan. In 1933 he left Italy due to opposition from certain milieus hostile to an independent *Corriere*. He then moved to Switzerland, living in Zurich or Lugano, where for years he had invested his fortune with success. He also continued his charitable activities in favour of institutions and individuals.

After returning to Italy in 1950, Eugenio Balzan died in Lugano, Switzerland, on 15 July 1953.

Today, the Balzan Foundation, international in character and scope, acts jointly through two Foundations: one under Italian law and the other under Swiss law.

In Milan, the *International E. Balzan Prize Foundation – "Prize"* has the aim to promote, throughout the world, culture, science, and the most meritorious initiatives in the cause of humanity, peace and brotherhood among peoples, regardless of nationality, race or creed. This aim is attained through the annual award of four prizes in two general fields: literature, the moral sciences and the arts; medicine and the physical, mathematical and natural sciences.

Nominations for the prizes in the scientific and humanistic fields are received at the Foundation's request from the world's leading learned bodies. Candidates are selected by the *General Prize Committee*, composed of eminent European scholars and scientists. Since 2001, each prize is worth one million Swiss francs (about 620,000), half of which the prizewinner must destine for research work, preferably involving young researchers.

At intervals of not less than three years, the Balzan Foundation also awards a "Prize for Humanity, Peace and Brotherhood among Peoples", of varying amounts.

In Zurich, the *International E. Balzan Prize Foundation – "Fund"* administers Eugenio Balzan's estate.

Balzan Prizewinners for Literature, Moral Sciences, and the Arts; for Physical, Mathematical and Natural Sciences, and Medicine

2008
Wallace S. Broecker (USA) The Science of Climate Change
Maurizio Calvesi (Italy) The Visual Arts since 1700
Ian H. Frazer (Australia/UK) Preventive Medicine, including Vaccination
Thomas Nagel (USA/Serbia) Moral Philosophy

2007
Rosalyn Higgins (UK) International Law since 1945
Sumio Iijima (Japan) Nanoscience
Michel Zink (France) European Literature (1000–1500)
Jules Hoffmann (France) and Bruce Beutler (USA) Innate Immunity

2006
Ludwig Finscher (Germany) History of Western Music since 1600
Quentin Skinner (UK) Political Thought; History and Theory
Paolo de Bernardis (Italy) and Andrew Lange (USA) Observational Astronomy and Astrophysics
Elliot Meyerowitz (USA) and Christopher Somerville (USA/Canada) Plant Molecular Genetics

2005
Peter Hall (UK) The Social and Cultural History of Cities since the Beginning of the 16th Century
Lothar Ledderose (Germany) The History of the Art of Asia
Peter and Rosemary Grant (USA/UK) Population Biology
Russell Hemley (USA) and Ho-kwang Mao (USA/China) Mineral Physics

2004
Pierre Deligne (USA/Belgium) Mathematics
Nikki Ragozin Keddie (USA) The Islamic world from the end of the 19th to the end of the 20th century
Michael Marmot (USA) Epidemiology
Colin Renfrew (UK) Prehistoric Archaeology

2003
Reinhard Genzel (Germany) Infrared Astronomy
Eric Hobsbawm (UK/Egypt) European History since 1900

Wen-Hsiung Li (USA/Taiwan) Genetics and Evolution
Serge Moscovici (France/Romania) Social Psychology

2002
Walter Jakob Gehring (Switzerland) Developmental Biology
Anthony Thomas Grafton (USA) History of the Humanities
Xavier Le Pichon (France/Vietnam) Geology
Dominique Schnapper (France) Sociology

2001
James Sloss Ackerman (USA) History of Architecture (including town planning and landscape design)
Jean-Pierre Changeux (France) Cognitive Neurosciences
Marc Fumaroli (France) Literary History and Criticism (post 1500)
Claude Lorius (France) Climatology

2000
Ilkka Hanski (Finland) Ecological Sciences
Michel Mayor (Switzerland) Instrumentation and Techniques in Astronomy and Astrophysics
Michael Stolleis (Germany) Legal History since 1500
Martin Litchfield West (UK) Classical Antiquity

1999
Luigi Luca Cavalli-Sforza (USA/Italy) Science of Human Origins
John Elliott (UK) History, 1500–1800
Mikhael Gromov (France/Russia) Mathematics
Paul Ricœur (France) Philosophy

1998
Harmon Craig (USA) Geochemistry
Robert McCredie May (UK/Australia) Biodiversity
Andrzej Walicki (USA/Poland) The cultural and social history of the Slavonic world from the reign of Catherine the Great to the Russian revolutions of 1917

1997
Charles Coulston Gillispie (USA) History and Philosophy of Science
Thomas Wilson Meade (UK) Epidemiology
Stanley Jeyaraja Tambiah (USA/Sri Lanka) Social Sciences: Social Anthropology

1996
Arno Borst (Germany) History: Medieval Cultures
Arnt Eliassen (Norway) Meteorology
Stanley Hoffmann (France/USA/Austria) Political Sciences: Contemporary International Relations

1995
Yves Bonnefoy (France) Art History and Art Criticism (as applied to European Art from the Middle Ages to our times)

Carlo M. Cipolla (Italy) Economic History
Alan J. Heeger (USA) Science of New Non-Biological Materials

1994
Norberto Bobbio (Italy) Law and Political Science (governments and democracy)
René Couteaux (France) Biology (cell-structure with special reference to the nervous system)
Fred Hoyle (UK) and Martin Schwarzschild (USA/Germany) Astrophysics (evolution of stars)

1993
Wolfgang H. Berger (USA/Germany) Paleontology with special reference to Oceanography
Lothar Gall (Germany) History: Societies of the 19th and 20th centuries
Jean Leclant (France) Art and Archaeology of the Ancient World

1992
Armand Borel (USA/Switzerland) Mathematics
Giovanni Macchia (Italy) History and Criticism of Literatures
Ebrahim M. Samba (Gambia) Preventive Medicine

1991
György Ligeti (Austria/Hungary/Romania) Music
Vitorino Magalhães Godinho (Portugal) History: The emergence of Europe in the 15th and 16th centuries
John Maynard Smith (UK) Genetics and Evolution

1990
Walter Burkert (Germany) Study of the Ancient World (Mediterranean area)
James Freeman Gilbert (USA) Geophysics (solid earth)
Pierre Lalive d'Epinay (Switzerland) Private International Law

1989
Emmanuel Lévinas (France/Lithuania) Philosophy
Leo Pardi (Italy) Ethology
Martin John Rees (UK) High Energy Astrophysics

1988
Shmuel Noah Eisenstadt (Israel/Poland) Sociology
René Etiemble (France) Comparative Literature
Michael Evenari (Israel/France) and Otto Ludwig Lange (Germany) Applied Botany (including ecological aspects)

1987
Jerome Seymour Bruner (USA) Human Psychology
Richard W. Southern (UK) Medieval History
Phillip V. Tobias (South Africa) Physical Anthropology

1986
Otto Neugebauer (USA/Austria) History of Science

Roger Revelle (USA) Oceanography/Climatology
Jean Rivero (France) Basic Human Rights

1985
Ernst H.J. Gombrich (UK/Austria) History of Western Art
Jean-Pierre Serre (France) Mathematics

1984
Jan Hendrik Oort (Netherlands) Astrophysics
Jean Starobinski (Switzerland) History and Criticism of Literatures
Sewall Wright (USA) Genetics

1983
Francesco Gabrieli (Italy) Oriental Studies
Ernst Mayr (USA/Germany) Zoology
Edward Shils (USA) Sociology

1982
Jean-Baptiste Duroselle (France) Social Sciences
Massimo Pallottino (Italy) Sciences of Antiquity
Kenneth Vivian Thimann (USA/UK) Pure and Applied Botany

1981
Josef Pieper (Germany) Philosophy
Paul Reuter (France) International Public Law
Dan Peter McKenzie, Drummond Hoyle Matthews and Frederick John Vine (UK) Geology and Geophysics

1980
Enrico Bombieri (USA/Italy) Mathematics
Jorge Luis Borges (Argentina) Philology, Linguistics and Literary Criticism
Hassan Fathy (Egypt) Architecture and Urban Planning

1979
Torbjörn Caspersson (Sweden) Biology
Jean Piaget (Switzerland) Social and Political Sciences
Ernest Labrousse (France) and Giuseppe Tucci (Italy) History

1962
Paul Hindemith (Germany) Music
Andrej Kolmogorov (Russia) Mathematics
Samuel Eliot Morison (USA) History
Karl von Frisch (Austria) Biology

Balzan Prizewinners for Humanity, Peace and Brotherhood Among Peoples

2007
Karlheinz Böhm (Austria/Germany)

2004
Community of Sant'Egidio

2000
Abdul Sattar Edhi (Pakistan/India)

1996
International Committee of the Red Cross

1991
Abbé Pierre (France)

1986
United Nations Refugee Agency (UNHCR)

1978
Mother Teresa of Calcutta (India/Macedonia)

1962
H.H. John XXIII (Vatican City/Italy)

1961
Nobel Foundation

Name Index*

*Listed below are the names mentioned in the text, with the exception of the participants in the symposium.

MIX
Papier aus verantwortungsvollen Quellen
Paper from responsible sources
FSC® C105338

If you have any concerns about our products,
you can contact us on
ProductSafety@springernature.com

In case Publisher is established outside the EU,
the EU authorized representative is:
Springer Nature Customer Service Center GmbH
Europaplatz 3, 69115 Heidelberg, Germany

Printed by Libri Plureos GmbH
in Hamburg, Germany